Teacher's Book

simple modern maths 1

F C Boyde
formerly Head of Mathematics
Tower Hamlets School, London.

R A Court
Head of Mathematics
Stationers Company School.

with

A M Court

and

J C Hawdon

Nelson

Contents

unit 1 — **Revising Your Arithmetic**
Adding and taking away whole numbers, multiplying and dividing whole numbers, what are decimals? adding and taking away decimals.

unit 2 — **Different Kinds of Numbers**
Different number bases, changing numbers from base ten into other bases, binary numbers.

unit 3 — **Sets**
Sets, union and intersection, Venn diagrams.

unit 4 — **Number Patterns**
Finding the next number in a sequence.

unit 5 — **Angles**
Parallel lines and parallelograms, angles of a triangle, angles of a quadrilateral, drawing angles.

unit 6 — **Negative Numbers**
The number line, giving a meaning to negative numbers.

unit 7 — **More Arithmetic**
Multiplying and dividing decimals by whole numbers, multiplying and dividing decimals by decimals, multiplying and dividing numbers by 10, 100, 1000, cancelling fractions, changing fractions to decimals, finding fractions of numbers.

unit 8 — **Fixing Positions with Numbers**
Co-ordinates.

unit 9 — **Area and Volume**
Area of rectangle and triangle, area of other shapes, volume and surface area of rectangular block.

unit 10 — **Using Letters Instead of Numbers**
Formulae, making formulae simpler.

unit 11 — **Using Formulae**
Substituting values in formulae.

unit 12 — **Percentages**
What are percentages? writing one number as a percentage of another, finding percentages of numbers.

unit 13 — **Indices**
Positive whole number indices.

unit 14 — **Rectangular Numbers, Primes, and Factors**
Rectangular, composite, and prime numbers, finding prime numbers.

unit 15 **Number Pictures and Averages**
Pictograms, bar charts, averages, pie charts.

unit 16 **Equations**
Solving equations, using equations to solve problems.

unit 17 **Sharing Things Out**
Dividing quantities in a given ratio, simplifying ratios.

unit 18 **The Circle**
Finding the circumference of a circle, finding the area of a circle, practical problems.

unit 19 **What are the Chances?**
The meaning of probability, finding probabilities, finding frequencies given the probability. Coins and dice.

unit 20 **Graphs**
What are graphs? using graphs, drawing graphs.

unit 21 **Matrices**
What is a matrix? adding and taking away matrices, multiplying a matrix by a number.

unit 22 **Shapes that are the Same**
Congruent shapes, enlargement and similar figures, enlargements using co-ordinates, scale factors, finding the sides of similar figures.

To the Teacher

This book has been written for pupils who find mathematics difficult but who want to take CSE. SIMPLE MODERN MATHS, Books 1 and 2 together cover the main concepts required for the CSE examination. Many of the examples are very easy and the non-examination pupil will also find the books useful.

The examples have been carefully graded so that the first calculations in each exercise are very simple and introduce the new concept. They do not involve difficult arithmetic: for example if the exercise requires multiplication then the multiplier will be a one figure number. Practice in basic arithmetic is provided in Units 1 and 7.

As far as possible all the Units contain examples of the practical application of the topics so that pupils can see the relevance of mathematics to everyday life.

The solutions have been provided in most cases to four significant figures (the fourth figure has not been corrected up), and the authors have left it to the teacher to tell the pupil what degree of accuracy is required, and also at what stage to introduce the concept of rounding up.

For teachers who wish to follow their own syllabus, the book has been arranged so that the Units may be done in almost any order, and even the exercises within a Unit need not to be done in the suggested order. There are however a few occasions when certain Units are best done in the order given. For example the second part of Unit 8 (Negative co-ordinates) will require the work in Unit 6, and the algebra Units are best done in the order given.

It is recommended that pupils using this book be provided with exercise books ruled in $\frac{1}{2}$cm squares. This will remove the necessity of having to supply additional paper for graphical and statistical work, and for scale drawings.

unit 1 Revising Your Arithmetic

The first thing that you will be doing in this course will be to revise your arithmetic, and to use your arithmetic to solve simple practical problems.

Adding and taking away whole numbers

Exercise 1.1 Do these adding sums. For example:

2(b) is 8
        3 +

		a 2	b 3	c 7	d 11	e 23	f 49	g 83	h 184	i 237
1	5									
2	8									
3	9									
4	17									
5	25									
6	56									
7	97									
8	204									
9	732									
10	981									

11 357   12 46   13 784   14 276
    286   862   953   476
    267 +   208 +   96 +   804 +

15 4567   16 26 541
    2386   98 554
     539   3 492
    4789 +   134 678
                46 794 +

Do numbers 1 to 10 as taking-away sums. Always take the smaller number from the larger. For example:

3(d) is 11
          9 −

17 5468   18 3456   19 3478   20 2346
     345 −   1267 −   349 −   1297 −

21 45 678
    23 865 −

Exercise 1.2

In each of these examples you have to decide whether to add or take away.

1. A woman buys a loaf of bread at 8p, and a cake at 17p. How much did she have to pay?
2. A boy has 37p. If he spends 16p how much will he have left?
3. A boy buys a record costing 48p and a book costing 39p. How much does he have to pay?
4. If there are 37 in a class and 19 of them are girls, how many boys are there in the class?
5. A bottle of scent weighed 46 g when it was full, and the empty bottle weighed 27 g. What was the weight of scent in the bottle?
6. A coal dealer has 327 tonnes of coal in stock. If 243 tonnes are delivered how much coal has he now?
7. A woman buys a bottle of milk at 9p, a cake at 19p, and a tin of fruit at 21p. How much did she have to pay?
8. A girl has 93p. If she spends 36p on going to the cinema how much has she left?
9. What is the total cost of a packet of envelopes at 20p, a book of stamps at 25p, and a pad of writing paper at 13p?
10. A man buys a packet of grass seed at 23p, a box of fertilizer at 34p, and a sack of sand at 43p. What is the total cost?
11. If there are 135 boys in a school, and 143 girls, what is the total number of pupils in the school?
12. If there are 375 pupils in a school, and 146 of them are boys, how many girls are there?
13. A coal dealer has 456 tonnes of coal in stock; if he sells 264 tonnes how much has he left?
14. The number of breakfasts served in a hotel is 134, the number of lunches is 247, and the number of dinners served is 198. How many meals are served?
15. A car factory has 685 cars in stock. If 286 are sold how many are left?
16. A school has 143 boys in the first year, 153 in the second year, 147 in the third year, 132 in the fourth year, 97 in the fifth year, and 83 in the sixth year. How many boys are there in the school?
17. A man has £2568. He buys a car which costs him £1236 and a caravan which costs him £736. How much has he left?
18. Find the total cost of a car at £2463, a caravan at £956, and a boat at £753.

19 A man sells his house for £14 345 and he has £1475 in the bank. How much has he left if he buys another house costing £12 678 and a car costing £1792?

20 A man has £789 in the bank. He sells his car for £866 and buys another for £984. How much has he left in the bank?

Multiplying and dividing whole numbers

Exercise 1.3

Do these multiplication sums. Always put the smallest number underneath. For example:

$$4(b) \text{ is } \begin{array}{r} 24 \\ \underline{8} \times \end{array}$$

		a	b	c	d	e	f	g	h	i
		13	24	67	98	45	86	75	123	546
1	2									
2	5									
3	7									
4	8									
5	6									
6	29									
7	13									
8	41									
9	56									
10	791									

Do these division sums. If there is a remainder write it down as well.

	a	b	c	d	e
11	24 ÷ 3	16 ÷ 2	40 ÷ 5	35 ÷ 7	63 ÷ 9
12	49 ÷ 7	32 ÷ 8	18 ÷ 6	80 ÷ 8	15 ÷ 3
13	26 ÷ 3	17 ÷ 2	43 ÷ 5	37 ÷ 7	70 ÷ 9
14	51 ÷ 7	36 ÷ 8	21 ÷ 6	83 ÷ 8	17 ÷ 3
15	51 ÷ 3	92 ÷ 4	65 ÷ 5	75 ÷ 3	78 ÷ 6
16	104 ÷ 8	98 ÷ 7	162 ÷ 9	225 ÷ 9	192 ÷ 8
17	97 ÷ 4	80 ÷ 6	69 ÷ 4	95 ÷ 7	109 ÷ 8
18	234 ÷ 13	276 ÷ 11	253 ÷ 21	754 ÷ 22	603 ÷ 23
19	953 ÷ 46	974 ÷ 63	983 ÷ 45	983 ÷ 73	938 ÷ 82

	f	g	h	i
11	56 ÷ 8	72 ÷ 9	48 ÷ 6	45 ÷ 9
12	45 ÷ 5	21 ÷ 3	42 ÷ 7	81 ÷ 9
13	59 ÷ 8	75 ÷ 9	53 ÷ 6	47 ÷ 9
14	48 ÷ 5	22 ÷ 3	45 ÷ 7	85 ÷ 9
15	95 ÷ 5	68 ÷ 4	96 ÷ 6	91 ÷ 7
16	288 ÷ 9	232 ÷ 8	243 ÷ 9	216 ÷ 8
17	165 ÷ 9	197 ÷ 8	290 ÷ 9	230 ÷ 8
18	246 ÷ 17	654 ÷ 22	838 ÷ 32	847 ÷ 42
19	649 ÷ 18	923 ÷ 73	937 ÷ 91	468 ÷ 93
20	4356 ÷ 234			
21	8743 ÷ 696			
22	87 653 ÷ 854			
23	89 456 ÷ 1346			
24	997 543 ÷ 6539			

Exercise 1.4

In each of these examples you have to decide whether to multiply or divide.

1. Find the cost of 7 plates if they cost 23p each.
2. Three friends have coffee and cakes in a cafe. If the bill came to 84p how much did each one have to pay?
3. A case of tea bags contains 25 boxes. Each box contains 9 bags. How many tea bags are there in the case?
4. A van has to carry 9 sacks of sand. Each sack weighs 73 kg. What is the weight of sand carried by the van?
5. Eight coaches are available to take 256 boys to games. How many boys will there be on each coach?
6. A train has 9 carriages each holding 68 people. How many people does the train hold?
7. Seven cups cost 266p. What is the cost of one cup?
8. Nine people have lunch which costs them 93p each. How big was the total bill?
9. 155 books have to be placed in 5 boxes. How many books will there be in each box?
10. 105 tonnes of earth have to be carried in 7 trucks. What weight will each truck carry?
11. A train has 8 carriages and it can carry 472 people. How many people can each carriage hold?
12. The distance around a field is 232 m. The field is to be fenced with pieces of fencing 4 m long. How many pieces will be needed?

13 What is the total cost of 26 fence posts that cost 39p each?

14 A man buys 37 chairs for his shop for £518. How much each are they?

15 A boy has saved £10·92 for a holiday which will last 14 days. How much will he have to spend each day? (Change £10·92 to pence)

16 A man is taking his family abroad on a holiday. He estimates that he will spend £27 per day. If the holiday lasts 17 days how much will he need?

17 A firm is taking all of its 1620 workers on a coach trip. If each coach holds 36, how many coaches will be needed?

18 A school wants to buy 255 chairs for the school hall. If the chairs cost £19 each, what will be the total cost?

19 Every year a certain firm divides some of its profits among its workers. How much will each worker get if the firm has to divide £22 528 between 352 workers

20 What is the total cost of 245 houses at £8457 each?

What are decimals?

What is the length of the line AB?

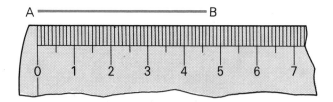

You will see that it is 4 cm long plus part of a cm.
Each cm is divided into ten parts, and the extra bit is six of these parts.

The length of the line is $4\frac{6}{10}$ cm. This is usually written 4·6 cm.

4·6 cm means the same as $4\frac{6}{10}$ cm

4·6 is called a decimal fraction or a **decimal**.

If you measure the line more accurately you might find that its length is $4\frac{62}{100}$ cm.

As a decimal this is written as **4·62**.

unit 1/page 9

If you measure something very accurately you will need to measure in thousandths of a cm. You might find that a length is $6\frac{235}{1000}$.

As a decimal this is written as **6·235** cm.

one figure after the point means tenths

two figures after the point means hundredths

three figures after the point means thousandths

Example

Write each of these fractions as decimals.

a $2\frac{5}{10} = 2·5$ d $14\frac{364}{1000} = 14·364$ g $\frac{3}{10} = 0·3$

b $8\frac{18}{100} = 8·18$ e $9\frac{48}{1000} = 9·048$ h $\frac{28}{100} = 0.28$

c $9\frac{7}{100} = 9·07$ f $3\frac{5}{1000} = 3·005$ i $\frac{372}{1000} = 0·372$

Exercise 1.5

Write the following fractions as decimals.

	a	b	c	d	e
1	$2\frac{3}{10}$	$4\frac{17}{100}$	$3\frac{1}{10}$	$5\frac{27}{100}$	$9\frac{2}{10}$
2	$12\frac{8}{10}$	$\frac{27}{100}$	$13\frac{7}{10}$	$\frac{237}{1000}$	$14\frac{462}{1000}$
3	$17\frac{9}{10}$	$\frac{6}{100}$	$20\frac{1}{10}$	$\frac{9}{100}$	$26\frac{6}{100}$
4	$4\frac{9}{1000}$	$9\frac{7}{100}$	$\frac{38}{1000}$	$127\frac{72}{100}$	$\frac{186}{1000}$

	f	g	h	i	j
1	$7\frac{5}{100}$	$6\frac{92}{100}$	$10\frac{6}{100}$	$8\frac{43}{100}$	$11\frac{7}{100}$
2	$\frac{6}{10}$	$15\frac{3}{10}$	$\frac{471}{1000}$	$\frac{5}{10}$	$16\frac{4}{10}$
3	$\frac{2}{10}$	$9\frac{3}{10}$	$\frac{39}{1000}$	$38\frac{42}{100}$	$41\frac{64}{1000}$
4	$8\frac{43}{1000}$	$\frac{271}{1000}$	$7\frac{27}{1000}$	$11\frac{6}{1000}$	$63\frac{42}{1000}$

Write the following decimals as fractions.

Example

$3·45 = 3\frac{45}{100}$

	a	b	c	d	e	f	g	h
5	3·2	5·37	4·8	9·64	9·4	9·05	9·64	67·05
6	18·2	0·75	19·8	0·287	85·543	0·6	2·5	0·635
7	19·7	0·03	43·7	0·3	53·06	0·9	6·9	0·056
8	9·004	9·04	0·053	149·83	0·764	3·95	0·345	6·023

Adding and taking away decimals

When adding and taking away decimals the numbers must be written so that the decimal points are underneath each other. Whole numbers such as 'seven' should be written as 7·0 or 7·00.

Example

Work out 5·6 + 2·3 and 19 + 3·6 + 0·367

```
  5·6          19·000
  2·3 +         3·600
  ───           0·367 +
  7·9          ───────
               22·967
```

You should notice that in the second example several 0's have been put in after '19'. This helps you to put the decimal point in the correct place in the answer.

Example

Work out 16·2 − 9·7 and 17 − 0·57

```
  16·2         17·00
   9·7 −        0·57 −
  ────         ──────
   6·5         16·43
```

Exercise 1.6

Do these adding sums.

For example **2c** is $\begin{array}{r}3·5\\9·0+\end{array}$

	a 1·3	b 2·7	c 9	d 16·5	e 0·6	f 0·45	g 0·05
1	1·6						
2	3·5						
3	3						
4	13·7						
5	0·7						
6	0·38						
7	0·02						
8	0·835						
9	13·8						
10	23·37						

11 23·3 + 18·56 + 0·56
12 0·987 + 23 + 18·67
13 238 + 9·007 + 189·465
14 56·865 + 0·098 + 24
15 89 + 8·7 + 0·0078
16 89·23 + 0·678 + 23 + 0·678

Do numbers 1 to 10 as take-away sums. Always write the smaller number at the bottom.

Exercise 1.7

In each of these examples you have to decide whether to add or take away.
When you add-up or take-away pence and £s you must either write the total amount as pence, or the total as £s.

Example

£2·35 + 67p.

You may do this as 2·35   or as 235
                           0·67 +   67 +
                           £3·02   302p = £3·02

1. Find the total cost of a book at £2·35 and a record at £1·47.
2. A boy has £1·56 and he spends £0·97 on a record. How much has he left?
3. Find the total cost of a book at £1·37 and a book at 57p.
4. A boy has £1·68 and he spends 83p on a record. How much has he left?
5. A roll of cloth contains 12·5 m. If 7·8 m are cut off how much is left?
6. A roll of cloth contains 14 m. If 3·9 m are cut off how much is left?
7. A man needs 4·6 m of water pipe for the bathroom, and 4·75 m of water pipe for the kitchen when he modernises his house. What is the total length that he needs?
8. A coal dealer has to supply 2·4 tonnes to one customer, 5·7 to another, and 6·5 to another. How many tonnes does he have to deliver?
9. A full bottle of oil weighs 2·45 kg, the empty bottle weighs 0·56 kg. What is the weight of oil in the bottle?
10. A truck weighs 2·56 tonnes. What will be the total weight of the truck when it carries 9·65 tonnes of sand?
11. A metal alloy contains 2·45 kg of zinc, and 0·56 kg of copper. What is the total weight of the alloy?
12. A tank contains 16·8 litres of wine. How much is left if 7·9 litres are taken out?
13. What is the total cost of eight litres of petrol for £0·73, and two litres of oil for £0·58?
14. A farmer has 14·7 m of fencing. How much has he left if he uses 9·8 m of the fencing?
15. How much change will a woman get from £2 if she spends £1·24 on meat, and £0·23 on vegetables?

16 Three friends have a meal in a cafe. Their bills are 53p, 78p, and 76p.
(a) What is the total bill?
(b) How much change will there be from £5?

17 A woman orders three lots of food from a shop costing £2·35, £0·89, and £7·93. She pays the shop £6 and then £2·50. How much does she still owe?

18 A shop receives four orders for plastic pipe. They are for 16 m, 4·5 m, 26·5 m, and 10 m. The shop has 20·5 m in stock and 30 m is due to be delivered the next day. How much more pipe will they need?

19 The weight of a bottle of salad oil is 1·23 kg, the weight of the empty bottle is 0·35 kg, and the weight of the stopper is 0·08 kg. What is the weight of the oil?

20 A bridge can carry a maximum weight of 29 tonnes. What is the maximum weight that a truck weighing 7·65 tonnes can carry over the bridge?

unit 2 Different Kinds of Numbers

Why do we write our numbers using the ten figures from 0 to 9?
It is obviously something to do with the fact that we have ten fingers.
What would happen if we had say eight or six fingers?
What would our numbers look like then?

If you count up to say 13 on your fingers the 1 stands for one set of ten, and the 3 stands for the extra numbers.

What would 13 look like if we only had eight fingers?
One way of writing it would be 15; which means one set of eight, and an extra five.

What would 9 look like if we only had six fingers?
One way of writing it would be 13 which means one set of six and three more.

You might say that the number of fingers is called the **base**.
We say that 13 is 15 in **base eight**.
$\quad$ We write $13 = 15_8$.
We say that 9 is 13 in **base six**.
$\quad$ We write $9 = 13_6$.

Example

Write 15 in base 9.
$15 = 9 + 6 = 16_9$ $\quad$ **one lot of nine plus six**.

Write 8 in base 5.
$8 = 5 + 3 = 13_5$ $\quad$ **one lot of five plus three**.

Write 23 in base 8.
$23 = 8 + 8 + 7 = 27_8$ $\quad$ **two lots of eight plus seven**.

Change 13_6 back to base ten (back to an ordinary number).
$13_6 =$ one lot of 6 plus $3 = 6 + 3 = 9$.

Change 37_9 back to base ten.
$37_9 =$ three lots of 9 plus $7 = 27 + 7 = 34$.

Exercise 2.1

Write these numbers in the base shown.

	a	b	c	d	e	f
1 Write 11 in base	4	5	6	7	8	9
2 Write 9 in base	4	5	6	7	8	9
3 Write 15 in base	4	5	6	7	8	9
4 Write 5 in base	4	5	6	7	8	9
5 Write 27 in base	4	5	6	7	8	9

Change these numbers back to base 10.

	a	b	c	d	e	f
6	12_5	18_9	14_6	14_7	10_8	12_4
7	13_6	12_5	13_7	15_9	16_7	11_8
8	23_5	28_9	36_8	31_5	42_6	21_7
9	32_5	63_7	38_9	41_5	37_8	68_9
10	78_9	37_8	65_7	42_5	33_5	71_9

Changing large numbers back into base ten

How do we change large numbers like 1463_8 back into base ten?

One way of doing sums like this is to look at the meaning of each of the figures in ordinary numbers.

Look at 3789_{10}.

This means 3 lots of 1000, 7 lots of 100, 8 lots of 10, and 9 units. Or we could write it like this:

$10 \times 10 \times 10 =$ 1000	$10 \times 10 =$ 100	10	1
3	7	8	9

If we only had eight fingers, that is if we were counting in base 8, the numbers at the top would be $8 \times 8 \times 8$, 8×8, 8, and 1.

We can now change 1463_8 into base ten as follows.

$8 \times 8 \times 8 =$ 512	$8 \times 8 =$ 64	8	1
1	4	6	3

1 lot of 512 = 512
4 lots of 64 = 256
6 lots of   8 = 48
3 lots of   1 =   3
                819

So we can say $1463_8 = 819_{10}$.

Example

Change 2354_6 to base 10.

$6 \times 6 \times 6 =$ 216	$6 \times 6 =$ 36	6	1
2	3	5	4

$2 \times 216 = 432$
$3 \times 36 = 108$
$5 \times 6   = 30$
$4 \times 1   =   4$
              574

$2354_6 = 574$.

Exercise 2.2

Use this table to change the following numbers to base ten.

5 × 5 × 5 = 125	5 × 5 = 25	5	1

1 3142_5 2 4231_5 3 1422_5 4 1421_5 5 2431_5

Use this table to change the following numbers to base ten.

7 × 7 × 7 = 343	7 × 7 = 49	7	1

6 5624_7 7 3442_7 8 6261_7 9 2361_7 10 5341_7

Change these numbers to base ten.

	a	b	c	d	e	f	g
11	2453_6	5431_6	254_6	3212_4	3321_4	2132_4	322_4
12	1211_3	2121_3	121_3	7354_8	2653_8	765_8	2674_8
13	5874_9	853_9	1017_9	3000_9	7534_8	1212_5	6754_8
14	1011_2	1100_2	1010_2	1111_2	1000_2	10101_2	111_2

Changing numbers in base ten into other bases

The best way to change a number in base ten to any other base is to keep dividing by the base and writing down the remainders. Writing down the remainders backwards will give the number in its new base.

Example

Change 819 to base 8.

8	819	R3	8 into 819 goes 102 and remainder 3
8	102	R6	8 into 102 goes 12 and remainder 6
8	12	R4	8 into 12 goes 1 and remainder 4
8	1	R1	8 into 1 goes 0 and remainder 1
	0		

$$819 = 1463_8$$

You will see that this is correct if you look at the example on page 14.

page 16/unit 2

Example

Change 574 to base 6.

```
6 | 574 | R4
6 |  95 | R5
6 |  15 | R3
6 |   2 | R2
```

$574 = 2354_6$

You will see that this is correct if you look at the example on page 14.

Exercise 2.3

Change these numbers...

		a	b	c	d	e
1	...to base 4	365	762	138	725	1784
2	...to base 7	298	594	284	285	8742
3	...to base 5	289	903	386	294	4390
4	...to base 6	429	854	394	729	7394
5	...to base 8	285	539	329	739	2973
6	...to base 2	23	47	16	41	243

Writing numbers in base two

In exercises 2.2 and 2.3 you saw that numbers written in base 2 only need two figures: 0 and 1.
Examples are $7 = 111_2$, $10 = 1010_2$, $43 = 101\,011_2$.
This is used in the design of computers because if numbers are stored in the memory of the computer in **binary** (base 2) it makes the design of the computer much simpler.

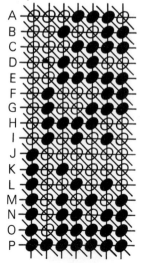

unit 2/page 17

One way of storing numbers is in a **magnetic core store**. In this a large number of metal rings are threaded on wires, and they can be magnetized by passing electric currents through the wires. If a ring is magnetized it can stand for 1 and if it is not magnetized it can stand for 0.

Example

A drawing of a **magnetic core store** is shown opposite. If each row stores one number, and if the dark rings represent the magnetised ones, what number would be stored in row A?

First we have to write the number in base 2 like this 1110, and then change it to base 10 as in Exercise 2.2

$2 \times 2 \times 2 \times 2 \times 2 =$ 32	$2 \times 2 \times 2 \times 2 =$ 16	$2 \times 2 \times 2 =$ 8
		1

$2 \times 2 =$ 4	2	1
1	1	0

$1110 = 8 + 4 + 2 = \mathbf{14}$

Exercise 2.4

1 Write down (in base 10) the numbers that are stored in lines B to P.

Use squared paper to show how each of the following numbers could be stored in a computer.

	a	b	c	d	e	f	g	h	i	j
2	13	17	22	27	30	33	34	43	47	52
3	61	65	67	73	87	103	117	120	123	140

Example

26. First we divide by 2 to find the remainders.

```
2 | 26   R0
2 | 13   R1
2 |  6   R0
2 |  3   R1
2 |  1   R1
     0
```

$26 = 11\,010_2$ (remember to write the remainder from the bottom).

Now mark the dots on the squared paper (there is no need to draw any lines).

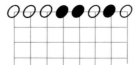

unit 3 Sets

In mathematics a collection of things is called a **set**. There are two ways of describing a set. We can either make a list of the things in the set, or we can describe the set in words.

Example

{Whole numbers from 1 to 5}   this may also be written   {1, 2, 3, 4, 5}.

{John, Eric, Peter, Simon}   this may also be written {The boys I cycle with}.

{a, b, c, d, e}   this may also be written as   {The first five letters of the alphabet}.

You should notice that sets are written in brackets like this, { } and that the things in the set, usually called the **elements**, are separated by commas.
It does not matter what order we put the elements in. The first set above could also be written {4, 2, 5, 1, 3}.

If you want to talk about a set you can use a single letter to stand for it. For example, we can write

$$N = \{1, 2, 3, 4, 5\}.$$

Example

Write down the set of whole numbers from 6 to 10. How many elements are there in this set?

The set is {6, 7, 8, 9, 10}.
There are five elements in the set.

Describe this set in words.

{a, b, c, d, e, f}.
{The first six letters of the alphabet}.

Exercise 3.1

Describe the following sets in words and state the number of elements in each set.

1 {Monday, Tuesday, Wednesday, Thursday, Friday, Saturday, Sunday}.
2 {z, y, x, w, v}.
3 {2, 4, 6, 8, 10}.
4 {f, g, h, i, j}.
5 {1, 3, 5, 7, 9}.
6 {January, February, March, April, May}.
7 {95, 96, 97, 98, 99, 100}.
8 {September, October, November, December}.

List the elements of the following sets

9 {The last five letters of the alphabet}.
10 {The whole numbers greater than 15 and less than 20}.
11 {Whole numbers from 20 to 25}.
12 {Months of the year beginning with J}.

13 {Odd numbers, less than 30, and divisible by 3}.
14 {The years from 1970 to 1980}.
15 {Whole numbers less than 10, but greater than 0}.

Putting sets together

In a class of boys there are some who go mountain walking (M), and some who go rock climbing (R). Let us call these two sets M and R.

$$M = \{Roy, Alan, Peter, Ron, Michael\}$$
$$R = \{John, Simon, Alan, Ron, Roy\}$$

If we want to make a list of boys who are interested either in rock climbing or in mountain walking (or in both) this is called the **union** of the two sets. It is written like this:

$$M \cup R = \{Roy, Alan, Peter, Ron, Michael, John, Simon\}$$

If we want to make a list of boys who are interested in rock climbing *and* in mountain walking this is called the **intersection** of the two sets. It is written like this:

$$M \cap R = \{Roy, Alan, Ron\}$$

Example

Copy and complete:
$$\{1, 2, 3, 4\} \cup \{3, 4, 5\} =$$
$$\{1, 2, 3, 4\} \cap \{3, 4, 5\} =$$
$$\{1, 2, 3, 4\} \cup \{3, 4, 5\} = \{1, 2, 3, 4, 5\}$$
$$\{1, 2, 3, 4\} \cap \{3, 4, 5\} = \{3, 4\}$$

If $X = \{p, q, r, s\}$ and $Y = \{q, s, t\}$ write down $X \cup Y$ and $X \cap Y$.

$$X \cup Y = \{p, q, r, s, t\} \qquad X \cap Y = \{q, s\}$$

If $P = \{1, 2, a, b\}$ and $Q = \{4, 5, c, d\}$ write down the intersection of P and Q.
Since there are no elements which are in P as well as in Q we have to leave the brackets empty and write
$$P \cap Q = \{ \quad \}$$

Exercise 3.2

Copy and complete the following:

1 $\{a, b, c, d\} \cup \{b, c, d, e\} =$
2 {Peter, Paul, Mary} $\cup$ {John, Paul, George, Ringo} =

$A = \{1, 3, 5, 7\}$, $B = \{2, 4, 6, 8, 10\}$, $C = \{2, 3, 5, 7\}$, $D = \{3, 6, 9\}$. Use this to complete the following:

3 $A \cup B =$
4 $A \cup C =$

5 $B \cup D =$
6 $\{a, b, c, d\} \cap \{b, c, d, e\} =$
7 $\{Peter, Paul, Mary\} \cap \{John, Paul, George, Ringo\} =$
8 $B \cap C =$
9 $A \cap D =$
10 $A \cap C =$
11 $A \cap B =$
12 $A \cup C \cup D =$
13 $B \cap C \cap D =$

In these questions do the part in brackets first:
14 $A \cup (B \cap C) =$
15 $B \cap (C \cup D) =$
16 $(B \cap C) \cup D =$

Venn diagrams

One way of making a list of the boys in the two sets on page 19 is as follows:

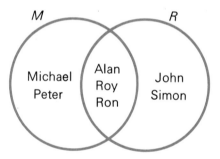

This is called a **Venn diagram** and it shows quite clearly who does what. For example we can see that John and Simon do rock climbing only, and Alan, Roy, and Ron are the only ones who do both rock climbing and mountain walking.

Example

$F = \{Boys\ who\ play\ football\} = \{Michael, Alan, John, Peter, Roy\}$
$C = \{Boys\ who\ play\ cricket\} = \{Peter, Roy, Simon, Ron\}$

a Show this on a Venn diagram.
b Write down the set of boys that play both cricket and football.
c Write down the set of boys that only play football and do not play cricket.
d Write down the set of boys that only play cricket and do not play football.

Example

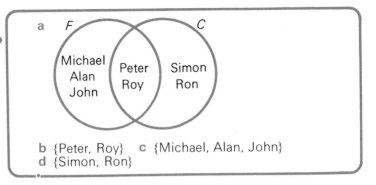

b {Peter, Roy} c {Michael, Alan, John}
d {Simon, Ron}

If you want to shade in the part of the Venn diagram that stands for C∩F it will look like this.

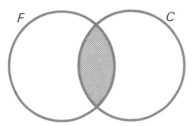

If you want to shade in the part of the Venn diagram that stands for C∪F it will look like this.

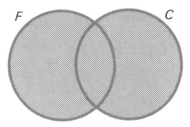

Exercise 3.3

1 C = {children who like chess},
   D = {children who like draughts}.

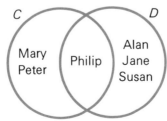

a Which children like chess but not draughts?
b Which children like draughts but not chess?
c Copy and complete this intersection: C∩D.

2 C={people who like coffee},
   T={people who like tea}.

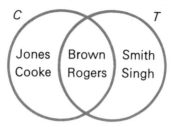

a List all the people who like tea.
b List all the people who like both tea and coffee.

3 F={children who like football},
   H={children who like hockey},
   B={children who like basketball}.

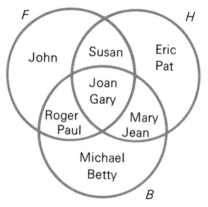

a Which children like hockey and basketball but not football?
b Who only likes football?
c Who likes all three games?

For the following questions draw Venn diagrams illustrating the information given.

4 M={children who like mathematics}={Roger, Simon, Peter, Susan},
   E={children who like English}={Roger, Mary, Anne, Susan}.
a Who likes mathematics and English?
b Who likes English but not mathematics?

5 E={even numbers between 1 and 11},
   S={numbers from 3 to 9}.

Copy and complete the following intersection: S∩E.

6 J={months of the year beginning with J},
   Y={months of the year ending with Y}.
a Copy and complete the intersection J∩Y.
b What months start with J and end with Y?

7 M = {letters in the word 'olympiad'},
   N = {letters in the word 'mineral'}.
  a List the letters that are in both words.
  b List the letters that are in 'olympiad' but not in 'mineral'.

8 T = {multiples of 3 between 1 and 31},
   E = {even numbers between 1 and 31},
   F = {multiples of 5 between 1 and 31}.
  a How many even multiples of 3 lie between 1 and 31?
  b Copy and complete: $(T \cup E) \cap F =$

9 O = {numbers between 0 and 100 ending in '1'},
   S = {multiples of 7 between 1 and 100},
   E = {multiples of 11 between 1 and 100}.
  Copy and complete: (a) $(O \cup S) \cap E =$
                     (b) $O \cup (S \cap E) =$.

More about Venn diagrams

Instead of naming the elements in the Venn diagram it is sometimes more useful to say how many elements there are.

Example

The Venn diagram shows how many people in a block of flats have vacuum cleaners and refrigerators.

a How many people have both?
b How many people have a vacuum cleaner?
c How many people have a refrigerator?
d How many people have a vacuum cleaner and not a refrigerator?
e How many people have a refrigerator and not a vacuum cleaner?
f How many people are included in the Venn diagram?

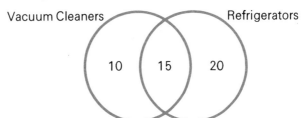

a 15 people have both.
b Number with vacuum cleaner = 10 + 15 = 25.
c Number with refrigerator = 15 + 20 = 35.
d Number with vacuum cleaner only = 10.
e Number with refrigerator only = 20.
f Number of people in diagram = 10 + 15 + 20 = 45.

Example

In a survey in a small town 55 people were asked if they had a radio or a television set. 30 said they had a television set, and 45 said they had a radio.

a Show this on a Venn diagram.
b How many had both?
c How many had only a television set?
d How many had only a radio?

Since 30+45=75 and only 55 people were asked, it is obvious that some had both. We can get the number who had both by taking 55 from 75 to give 20. The number with only a television set = 30−20 = 10. The number with only a radio = 45−20 = 25.

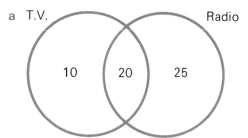

a

From the working or the diagram we can see that:
b 20 had both.
c 10 had only a television set.
d 25 had only a radio.

Exercise 3.4

1 The Venn diagram shows the number of boys in a class who play rugby and soccer.

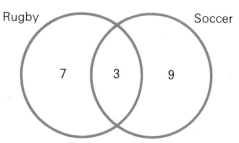

a How many boys play both rugby and soccer?
b How many boys play only soccer?
c How many boys play rugby?
d How many boys play either rugby or soccer, or both?

2 Some girls in a school were asked if they did cooking or needlework at home. The results of this inquiry are shown in the Venn diagram.

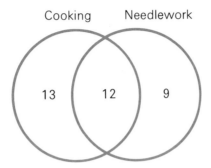

- **a** How many of the girls cook?
- **b** How many do needlework?
- **c** How many do both?
- **d** How many only do cooking?

3 A class of children were asked which hand they wrote with. Thirty of them could write with their right hand, five could write with their left hand and two with both hands. Draw a Venn diagram illustrating this information.

- **a** How many can write only with their right hand?
- **b** How many of them write only with their left hand?
- **c** How many children are there in the class?

4 In a youth club 20 teenagers said that they either went dancing or to the cinema regularly or both. Fifteen said they went to the cinema and 12 said they went dancing. Show this information by means of a Venn diagram.

- **a** How many went both dancing and to the cinema regularly?
- **b** How many went only to the cinema?
- **c** How many only went dancing?

5 A survey was made of the kind of books people borrowed from a library. The number of people reading science-fiction, westerns, and detective stories is shown in the diagram.

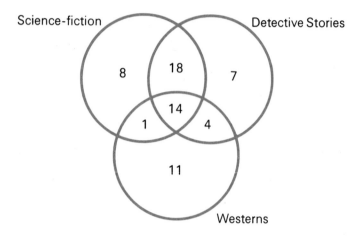

a How many read science-fiction?
b How many read westerns?
c How many read detective stories?
d How many read all three?
e How many read science-fiction and detective stories but not westerns?
f How many read detective stories and westerns but not science-fiction?

6 The Venn diagram in this question shows the numbers in the survey who read romances, historical novels and humorous books.

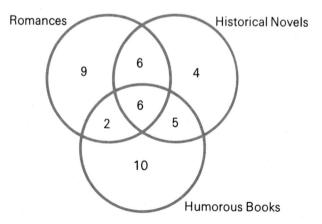

a How many read at least two kinds of books?
b How many read only one kind?

c Which, of romances, historical novels and humorous books, are the most popular?

d How many read both romances and historical novels?

e How many read both humorous books and historical novels but not romances?

For the following questions draw Venn diagrams to illustrate the information given.

7 In a class of 30 children, 18 like tennis, 17 like swimming and 15 like cricket. Eight children like both tennis and cricket, 9 like both tennis and swimming, 6 like both cricket and swimming, and 5 like all three.

a How many like only cricket?

b How many like only swimming?

c How many like only tennis?

d How many like tennis and cricket but not swimming?

8 In a survey amongst 100 people, 23 smoke cigarettes, 24 smoke cigars, 21 smoke a pipe, 7 smoke a pipe and cigarettes, 11 smoke cigars and cigarettes, 12 smoke cigars and a pipe and 4 smoke all three.

a How many smoke only cigarettes?

b How many smoke only a pipe?

c How many do not smoke?

9 Forty people bought newspapers, 19 bought the *Daily Review*, 19 the *Daily Report* and 21 the *Daily Chronicle*. Of those who bought two or more papers eight bought the *Review* and the *Chronicle*, ten the *Report* and the *Chronicle* and four the *Review* and the *Report*.

a How many bought all three papers?

b How many bought only the *Daily Chronicle*?

c How many bought the *Daily Review* and the *Daily Report* but not the *Daily Chronicle*?

unit 4 Number Patterns

What is the next number in this sequence? 1, 2, 3, 4, 5, 6, ?
What is the next number in this sequence? 3, 5, 7, 9, 11, 13?

The answers are obviously 7 and 15. In the first set of numbers each number is **one** more than the number in front of it. In the second set each number is **two** more than the number in front of it.

Example

Find the next two numbers in each of these sequences.
a 7, 10, 13, 16,
b 25, 21, 17, 13,
c 2, 2½, 3, 3½, 4,

a In this case each number is 3 more than the number before. The next numbers are 19 and 22.
b In this case each number is 4 less than the number before. The next numbers are 9 and 5.
c In this case each number is ½ more than the number in front. The next two numbers are 4½ and 5.

What is the next number in this sequence? 2, 4, 8, 16,
What is the next number in this sequence? 1, 3, 9, 27,
In the first set each number is twice the number in front of it. The answer is 32. In the second set each number is three times the number in front of it. The next number is $3 \times 27 = 81$.

Example

Find the next two numbers in each of these sequences.
a 3, 6, 12, 24,
b 64, 32, 16, 8,

a In this case each number is twice the number in front of it. The next numbers are 48 and 96.
b In this case each number is half the number in front of it. The next numbers are 4 and 2.

Exercise 4.1

Find the next two numbers in each of these sequences.

a
1 2, 4, 6, 8,
2 47, 45, 43, 41,
3 14, 17, 20, 23,
4 1, 6, 11, 16,
5 17, 21, 25, 29,
6 2, 8, 14, 20,
7 29, 25, 21, 17,
8 5, 13, 21, 29,
9 6½, 6, 5½, 5,
10 7·1, 7·2, 7·3, 7·4,

b
1, 2, 4, 8,
4, 8, 16, 32,
5, 10, 20, 40,
2, 6, 18, 54,
6, 12, 24, 48,
729, 243, 81, 27,
5, 15, 45, 135,
972, 324, 108, 36,
$\frac{1}{4}, \frac{1}{2}, 1, 2,$
$\frac{1}{32}, \frac{1}{16}, \frac{1}{8}, \frac{1}{4},$

11 1·1, 2·2, 4·4, 8·8,
12 7, 13, 19, 25,
13 7·1, 14·2, 28·4, 56·8,
14 256, 128, 64, 32,
15 3, 12, 48, 192,
16 6·1, 5·3, 4·5, 3·7,

47·5, 47·1, 46·7, 46·3,
$\frac{1}{243}, \frac{1}{81}, \frac{1}{27}, \frac{1}{9},$
$\frac{1}{625}, \frac{1}{125}, \frac{1}{25}, \frac{1}{5},$
100, 10, 1, 0·1,
1215, 405, 135, 45,
$1\frac{1}{3}, 2, 2\frac{2}{3}, 3\frac{1}{3}$

More number patterns

Look at this set of numbers. Can you find the next two?

2, 5, 10, 17, 26,

If you tried the two methods that we have used before you will find that neither of them works. For example 5 is 3 more than 2, but 10 is not 3 more than 5. Also 10 is twice 5, but 17 is not twice 10.

Is there any pattern to this set of numbers? Let us write down the numbers and then write down how much more each number is than the one in front of it.

```
2     5     10     17     26
   3     5      7      9
```

You should see that there is a pattern to the second set of numbers so we can write:

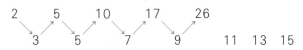

If you follow the arrows along you will see that you can get each number on the top line by adding the two numbers in front of it.

$2+3=5; \quad 5+5=10; \quad 10+7=17; \quad 17+9=26$

If we continue like this we get

$26+11=37; \quad 37+13=50; \quad 50+15=65$

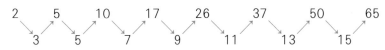

Example

Find the next three numbers in this sequence.
2, 6, 12, 20, 30,
If you use the arrow method you will see that the next three numbers are 42, 56, 72.

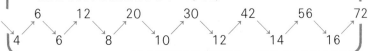

page 30 / unit 4

Exercise 4.2

Find the next two numbers in each of these sequences.

a
1. 1, 2, 4, 7, 11,
2. 3, 4, 6, 9, 13,
3. 1, 3, 7, 13, 21,
4. 82, 80, 76, 70, 62,
5. 1, 4, 10, 19, 31,
6. 43, 40, 35, 28,
7. 2, 4, 9, 17, 28,
8. 1, 20, 36, 49, 59,
9. 1·1, 11·1, 31·1, 61·1, 101·1,
10. $2\frac{3}{4}$, $3\frac{1}{2}$, $4\frac{1}{2}$, $5\frac{3}{4}$, $7\frac{1}{4}$,

b
1. 2, 7, 13, 20, 28,
2. 3, 7, 12, 18, 25,
3. 27, 26, 24, 21, 17,
4. 1, 7, 15, 25, 37,
5. 3, 6, 11, 18, 27,
6. 2, 7, 17, 32, 52,
7. 63, 50, 39, 30, 23,
8. 1, 11, 26, 46, 71,
9. $1\frac{1}{2}$, 2, 3, $4\frac{1}{2}$, $6\frac{1}{2}$,
10. 3·4, 4·3, 5·5, 7·0, 8·8,

Exercise 4.3

1 Look at the lines underneath. Copy and complete lines **e, f, g**.

- **a** 1   = 1 = 1 × 1
- **b** 1 + 3   = 4 = 2 × 2
- **c** 1 + 3 + 5   = 9 = 3 × 3
- **d** 1 + 3 + 5 + 7   = 16 = 4 × 4
- **e** 1 + 3 + 5 + 7 + 9   = 25 = — × —
- **f** 1 + 3 + 5 + 7 + 9 + 11   = 36 = — × —
- **g** 1 + 3 + 5 + 7 + 9 + 11 + 13 = — = — × —

You have been learning to add up the odd numbers.
- **h** Add up the odd numbers from 1 to 15.
- **i** Add up the odd numbers from 1 to 17.

Copy and complete these lines.
- **j** The sum of the first 2 odd numbers = 2 × 2 =
- **k** The sum of the first 3 odd numbers = 3 × 3 =
- **l** The sum of the first 4 odd numbers = — × — =
- **m** The sum of the first 8 odd numbers = — × — =
- **n** The sum of the first 10 odd numbers = — × — =
- **o** The sum of the first 15 odd numbers = — × — =

2 Look at the lines underneath. Copy and complete lines **e, f, g**.

- **a** 2   = 2 = 1 × 2
- **b** 2 + 4   = 6 = 2 × 3
- **c** 2 + 4 + 6   = 12 = 3 × 4
- **d** 2 + 4 + 6 + 8 = 20 = 4 × 5

e $2+4+6+8+10$ $=30=-\times-$
f $2+4+6+8+10+12$ $=42=-\times-$
g $2+4+6+8+10+12+14=—=-\times-$

You have been learning to add up the even numbers.
h Add up the even numbers from 2 to 16.
i Add up the even numbers from 2 to 18.

Copy and complete these lines.
j The sum of the first 2 even numbers $=2\times3=-$
k The sum of the first 3 even numbers $=3\times4=-$
l The sum of the first 4 even numbers $=-\times-=-$
m The sum of the first 8 even numbers $=-\times-=-$
n The sum of the first 10 even numbers $=-\times-=-$
o The sum of the first 15 even numbers $=-\times-=-$

3 Look at the lines underneath. Copy and complete lines e, f, g.
a 1 $= 1 = \frac{1}{2}\times\ 2 = \frac{1}{2}\times 1\times 2$
b $1+2$ $= 3 = \frac{1}{2}\times\ 6 = \frac{1}{2}\times 2\times 3$
c $1+2+3$ $= 6 = \frac{1}{2}\times 12 = \frac{1}{2}\times 3\times 4$
d $1+2+3+4$ $=10 = \frac{1}{2}\times 20 = \frac{1}{2}\times 4\times 5$
e $1+2+3+4+5$ $=15 = \frac{1}{2}\times 30 = \frac{1}{2}\times-\times-$
f $1+2+3+4+5+6$ $=21 = \frac{1}{2}\times—=\frac{1}{2}\times-\times-$
g $1+2+3+4+5+6+7=—=\frac{1}{2}\times—=\frac{1}{2}\times-\times-$

You have been learning to add up the whole numbers.
h Add up the whole numbers from 1 to 8.
i Add up the whole numbers from 1 to 9.

Copy and complete these lines.
j The sum of the first 2 whole numbers
   $=\frac{1}{2}\times 2\times 3 = \frac{1}{2}\times\ 6 =-$
k The sum of the first 3 whole numbers
   $=\frac{1}{2}\times 3\times 4 = \frac{1}{2}\times 12 =-$
l The sum of the first 4 whole numbers
   $=\frac{1}{2}\times 4\times 5 = \frac{1}{2}\times 20 =-$
m The sum of the first 8 whole numbers
   $=\frac{1}{2}\times-\times-=\frac{1}{2}\times-=-$
n The sum of the first 10 whole numbers
   $=-\times-\times-=-\times-=-$
o The sum of the first 15 whole numbers
   $=-\times-\times-=-\times-=-$

unit 5 Angles

You should know that in a complete rotation the angle turned through is 360 degrees, usually written 360°. This is shown below together with some other angles.

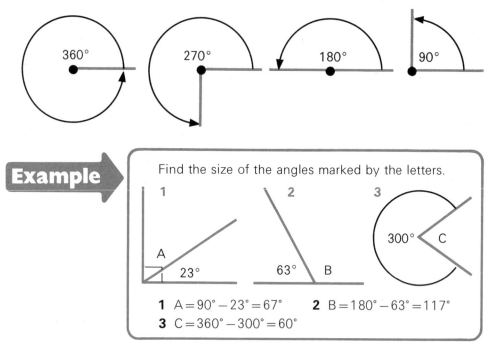

Example Find the size of the angles marked by the letters.

1 A = 90° − 23° = 67°   2 B = 180° − 63° = 117°
3 C = 360° − 300° = 60°

Exercise 5.1

Find the size of the angles marked by the letters.

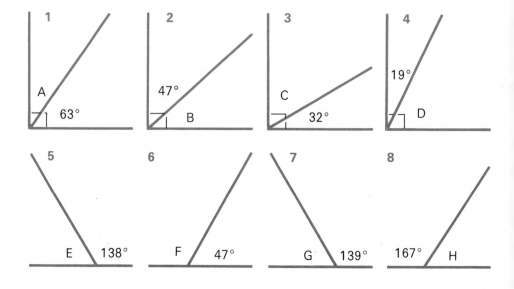

unit 5/page 33

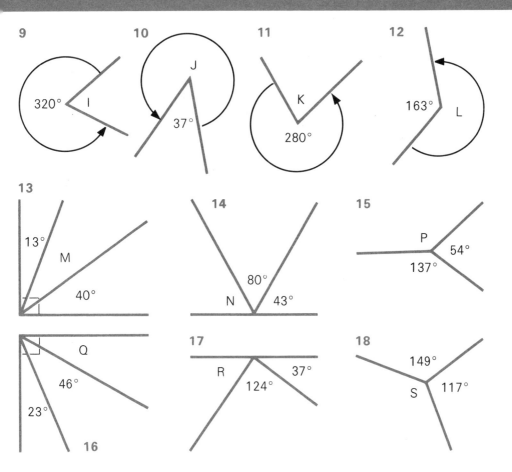

Parallel lines and parallelograms

Parallel straight lines are two or more straight lines that never meet however far they are drawn. Two sets of parallel lines are shown below. Usually parallel lines are marked with arrows to show you that they are parallel.

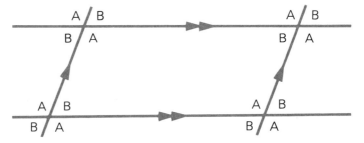

If you measure the angles you will find that all the angles marked with the letter A have the same size, and all the angles marked with the letter B have the same size.

page 34 / unit 5

When two pairs of parallel lines cross we get a four sided shape called a **parallelogram**. If you measure the sides and angles of a parallelogram you will find that the sides and angles marked with the same letters have the same size. An example of a parallelogram is given below.

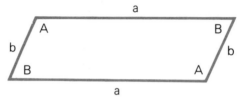

If you look at the two previous drawings you will see that A+B=180°.

Example

Find the size of the angles marked by the letters.

1 A = 180° − 47° = 133°
   B = 47°   C = A = 133°

2 B = 180° − 125° = 55°
   A = 125°

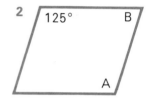

Exercise 5.2

Find the size of the angles marked by the letters.

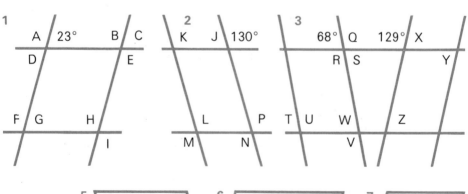

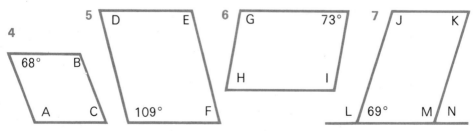

The angles of a triangle

Exercise 5.3

In each of these triangles measure the angles A, B, C, and then work out A + B + C.

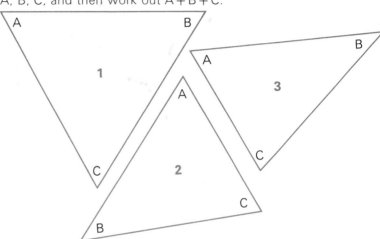

Draw some other triangles, measure their corner angles, and add them up. If you measured each angle correctly you should find that A + B + C adds up to about 180°. It can be shown that if each angle is measured exactly then we will always find that A + B + C = 180°.

the angles of a triangle add up to 180°

We can use this to find one of the angles of a triangle if we are given the other two.

Example

Find the size of angle A.

62° + 53° = 115°

A = 180° − 115° = 65°

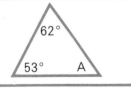

Find the size of the angle marked by the letters in these triangles.

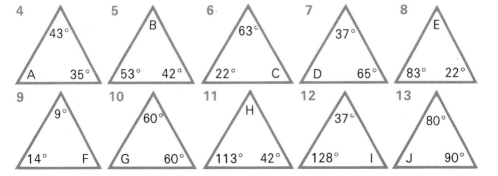

The angles of a quadrilateral

A shape with four straight sides is called a quadrilateral. Examples of quadrilaterals are given in Exercise 5.4.

Exercise 5.4

In each of these quadrilaterals measure the angles A, B, C, D and then work out A+B+C+D.

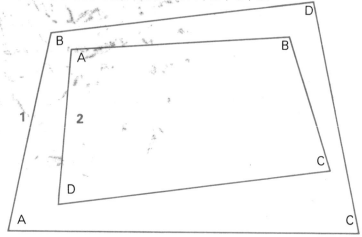

Draw some other quadrilaterals, measure their corner angles, and add them up.

If you measured each angle correctly you should find that A+B+C+D adds up to about 360°. It can be shown that if each angle is measured exactly then we will always find that A+B+C+D=360°.

the angles of a quadrilateral add up to 360°

We can use this to find one angle of a quadrilateral if we are given the other three.

Example

Find the size of angle A.

93°+107°+42°=242°

A=360°−242°=118°

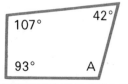

Find the size of the angle marked by the letter in these quadrilaterals.

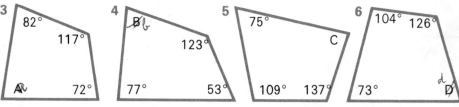

unit 5/page 37

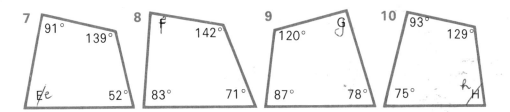

Drawing angles

So far in this Unit you have been measuring angles. The next exercise will give you practice in drawing angles.

Example

Make an accurate copy of the diagram below and measure the distance from A to B to the nearest tenth of a centimetre (0·1 cm).

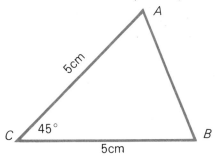

First you will need to draw an angle of 45°. Make the two lines longer than 5 cm. Then mark the points A and B, and then measure the distance from A to B. Your diagram should look like this.

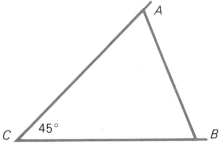

If you measure AB you will find that AB = 3·8 cm.

Exercise 5.5

Repeat the example above with the following angles.

1 23°	2 32°	3 67°	4 84°	5 63°
6 14°	7 53°	8 49°	9 38°	10 93°
11 107°	12 124°	13 153°	14 164°	15 173°
16 200°	17 243°	18 279°	19 305°	20 345°

unit 6 Negative Numbers

What is the answer to this sum? 7−5=
The answer is of course 2.

What is the answer to this sum? 5−7=?
You will probably say that it cannot be done because 7 is bigger than 5, but there are many occasions when an answer to a sum like this has to be found.

If you look at a thermometer you will see that as well as the numbers above 0, there are numbers below 0. These are called negative or minus numbers and are written like this; ⁻8. On a very cold day the weather man on television might say that the lowest temperature on that day was 8 degrees below zero, or minus 8 degrees.

If you take the numbers on a thermometer and place them on a line like this it is called a number line. We will use a number line to give a meaning to sums like 5−7.

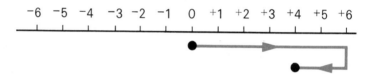

We could use a number line to do sums like 6−2. Start at 0 and move 6 spaces to the right, then move 2 spaces back. You will see that you end at 4 which is the correct answer. This has been done on the number line above.

We will now use the number line to work out the answer to 5−7.

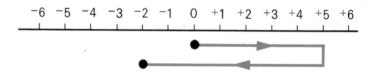

Start at 0 and move 5 spaces to the right, then move 7 spaces back. You will see that we end at ⁻2. So we can write 5−7=⁻2.

Here is another example. 4−9=⁻5.

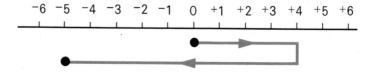

Exercise 6.1

Use this number line to work out the answers to these sums.

```
-10 -9 -8 -7 -6 -5 -4 -3 -2 -1  0 +1 +2 +3 +4 +5 +6 +7 +8 +9 +10
```

	a	b	c	d	e
1	2−6	6−9	8−10	1−3	2−7
2	8−9	1−1	2−9	1−2	4−3
3	7−8	8−3	1−4	10−6	1−6
4	4−11	9−13	10−12	9−14	7−10

	f	g
	4−6	7−3
	7−9	1−5
	4−12	8−12
	9−15	10−17

Example

Use the number line to see if you can get the answer to this.

$$6-8+4-9$$

Move 6 to the right, then move 8 to the left, then move 4 to the right, then move 9 to the left. You should end up at $^-7$.

See if you can get these answers using the number line. $6-10+2 = {^-2}$
$8-7-9 = {^-8}$

	a	b	c	d
5	3−6+1	10−5−8	7+1−9	5+3−12
6	2−4−4	9−3−4	1−2−6	0−4−5
7	8−13+6	9−12−6	10−17+5	1−5−3
8	10−12+9−4	3−7−2+6	9−2−7−6	9−5+1−5

Giving a meaning to negative numbers
Temperatures

If you are told that the temperature is 7°C and that it will fall 10°C we say that the new temperature is $7-10 = {^-3}$; the temperature will be $^-3$°C or 3°C below zero.

One use of negative numbers is to tell you when temperatures are below freezing.

page 40/unit 6

Bank balances

If you have £10 in the bank and take out £8 you have £2 left.

If you have £8 in the bank and take out £10 you owe the bank £2, or you have an **overdraft** of £2.

If you work out 8−10 on the number line you will get 8−10 = ⁻2.

A negative amount of money in the bank means that you have an overdraft, or you owe the bank some money.

Profit and loss

A club runs a dance. The band costs £12 and £17 worth of tickets are sold. What was the profit?
Profit = £17 − £12 = £5.

If the organizers only sold £8 worth of tickets their profit would be £8 − £12 = ⁻£4. They have obviously made a loss.

A negative profit means a loss.

Example
A man puts £7 in the bank, the next day he takes out £16, and the day after he puts in £5. How much has he got in the bank?
$$£7 - £16 + £5 = {}^-£4$$
He has an overdraft of £4.

Example
A club runs a dance. The organizers pay £8 for the hire of the hall, they take in £17 from tickets, but they have to pay the band £12. What profit do they make?
The money remaining is: $-£8 + £17 - £12 = ?$
To do this you will have to start by going *back* 8 on the number line, then 17 to the right, then back 12. You should end up at ⁻3.
The club made a loss of £3

Exercise 6.2

Do these examples using the number line and check that you get the correct answers.

$-9 + 2 = {}^-7$ (9 spaces back, and then 2 to the right)

$-7 - 3 + 9 = {}^-1$ (7 spaces back, then 3 spaces back, and then 9 spaces to the right)

$-4 - 3 = {}^-7$ (4 spaces back, and then 3 spaces back)

$-8 + 4 - 3 = {}^-7$

$-6 - 3 + 4 = {}^-5$

$-2 - 5 + 16 = 9$

	a	b	c
1	$-7+3$	$-7-3$	$-3+7$
2	$-5+4$	$-4-5$	$-4+5$
3	$-9+5+2$	$-10+6+4$	$-8+5+7$
4	$-1-2-3$	$-2-3-4+8$	$-7+6-6+9$

How much is left in the bank in the following?

5 A man has £7 in the bank. He takes out £13.
6 A man has £9 in the bank. He takes out £17.
7 A man has £8 in the bank. He takes out £5.
8 A man has nothing in the bank. He takes out £4.

A man starts with nothing in the bank. What is his final balance in the following examples?

9 £9 in, £12 out, £8 in, £10 out.
10 £7 in, £11 out, £9 in, £12 out.
11 £3 out, £4 out, £2 out.
12 £6 out, £3 out, £9 in, £7 in.
13 £6 out, £8 in, £5 in, £7 out.
14 £7 out, £12 out, £4 in, £8 in.

A club runs a dance. Work out the profit or loss in each of the following examples.

	Hire of hall	Sale of tickets	Hire of band
15	£7	£14	£9
16	£8	£12	£10
17	£3	£18	£11
18	£9	£12	£7
19	£17	£26	£21
20	£10	£31	£13
21	£5·20	£10·50	£6·40
22	£8·45	£37·57	£12·74

unit 7 More Arithmetic

Multiplying and dividing decimals by whole numbers

When you multiply and divide decimals by whole numbers you must remember to put the point in the answer above or below the point in the decimal.

Example

a $2 \cdot 16 \times 7 = 15 \cdot 12$
```
   2·16
      7 ×
  -----
  15·12
```

b $3 \cdot 25 \times 27 = 87 \cdot 75$
```
   3·25
     27 ×
  -----
  65·00
  22·75
  -----
  87·75
```

c $0 \cdot 67 \times 9 = 6 \cdot 03$
```
   0·67
      9 ×
  -----
   6·03
```

d $24 \cdot 9 \div 3 = 8 \cdot 3$
```
      8·3
   3 )24·9
```

e $23 \cdot 5 \div 7 = 3 \cdot 35$
```
       3·35
   7 )23·50
      21
      --
       25
       21
       --
        40
        35
```

f $0 \cdot 68 \div 12 = 0 \cdot 0566$
```
        0·0566
   12 )0·6800
       60
       --
        80
        72
        --
         80
         72
```

You will notice that in **e** and **f** the answers do not come out exactly. In this case we add one or two 0's to the end of the number we are dividing into so that we can get our answers to three figures, which is accurate enough for most purposes.

Exercise 7.1

Do these multiplication sums. For example, **3a** is 9×1.6.

		a 1·6	b 2·7	c 0·9	d 0·3	e 2·35	f 4·71	g 23·8
1	3							
2	6							
3	9							
4	5							
5	11							
6	23							
7	62							
8	82							
9	79							
10	243							

Do these dividing sums.

	a	b	c
11	$2.8 \div 4$	$3.9 \div 3$	$6.4 \div 8$
12	$12.4 \div 2$	$0.86 \div 2$	$0.65 \div 5$
13	$34.6 \div 3$	$25.6 \div 4$	$7.45 \div 5$
14	$0.78 \div 4$	$0.39 \div 7$	$0.693 \div 4$
15	$34.56 \div 12$	$17.75 \div 30$	$74.95 \div 62$
16	$24.57 \div 48$	$9.056 \div 56$	$976.43 \div 94$
17	$45.67 \div 95$	$356.89 \div 52$	$67.009 \div 32$
18	$56.94 \div 132$	$67.876 \div 356$	$989.4 \div 325$

	d	e	f
11	$5.4 \div 3$	$6.9 \div 3$	$4.8 \div 6$
12	$8.45 \div 5$	$56.8 \div 4$	$23.4 \div 9$
13	$9.73 \div 7$	$14.78 \div 4$	$9.62 \div 3$
14	$13.6 \div 7$	$18.8 \div 9$	$17.9 \div 3$
15	$96.47 \div 32$	$29.67 \div 40$	$0.78 \div 34$
16	$87.23 \div 23$	$679.34 \div 25$	$9.067 \div 34$
17	$67.98 \div 62$	$0.765 \div 24$	$0.964 \div 24$
18	$267.76 \div 854$	$8.098 \div 452$	$0.0765 \div 234$

g

11	$2.2 \div 2$	15	$0.987 \div 19$
12	$3.24 \div 6$	16	$234.56 \div 36$
13	$84.8 \div 9$	17	$56.907 \div 81$
14	$29.56 \div 7$	18	$987.567 \div 2765$

Exercise 7.2

In each of these examples you have to decide whether to multiply or divide.

1 Find the cost of 5 fence posts which cost £1·24 each.
2 Three friends have a meal which costs them £1·92. How much each did they have to pay?
3 What is the total weight of 7 blocks of stone, each of which weighs 1·56 tonnes?
4 Find the cost of 8 litres of petrol at £0·36 per litre.
5 Seven people hire a mini coach which costs them £9·52. How much each will they have to pay?
6 Eight dinner plates cost £4·56. What is the cost of one plate?
7 Five people have a set-price lunch which costs them £1·23 each. What will be the total cost of the lunch?
8 Find the cost of 6 cushions if each cushion costs £1·34.
9 44·8 tonnes of earth have to be carried in 8 trucks. How much will each truck have to carry?
10 Seven lengths of fencing have to be put up in a distance of 6·51 m. What is the length of each piece of fencing?
11 Find the cost of 16 litres of petrol at £0·38 per litre.
12 Six men win £152·73 between them on the pools. How much will they each get?
13 27 children are going on a day trip by train. The tickets cost £1·25 each. What will be the total cost?
14 A boy has saved £8·95 for a holiday which lasts five days. How much will he have to spend each day?
15 A man and his wife decide to spend 13 days at a hotel. The daily charge for both of them is £6·97. How much will they pay altogether?
16 The total cost of a day at the seaside for 23 boys and girls is £21·90. How much will they each have to pay?
17 A greengrocer buys 56 boxes of strawberries. He decides that he will sell them for a total of £13·50. How much will he charge per box?
18 29 rolls of cloth each hold 23·7 m of cloth. What is the total length of cloth on all the rolls?

Multiplying and dividing by decimals

Decimals are multiplied as follows.

Example $1 \cdot 9 \times 0 \cdot 7$

1 Remove the points and 0's at the beginning of each number. 19×7
2 Multiply the two numbers. $19 \times 7 = 133$
3 Count the number of figures after the point in the numbers you are multiplying and add them. $1 + 1 = 2$
4 Put this number of figures after the point in your answer. **1·33**

Example

a $1 \cdot 4 \times 4 \cdot 8 = \mathbf{6 \cdot 72}$
   ($14 \times 48 = 672$ Number of figures after point $= 1 + 1 = 2$).

b $1 \cdot 23 \times 0 \cdot 7 = \mathbf{0 \cdot 861}$
   ($123 \times 7 = 861$ Number of figures after point $= 2 + 1 = 3$).

c $0 \cdot 7 \times 0 \cdot 3 = \mathbf{0 \cdot 21}$
   ($7 \times 3 = 21$ Number of figures after point $= 1 + 1 = 2$).

d $0 \cdot 50 \times 2 \cdot 6 = \mathbf{1 \cdot 300}$
   ($50 \times 26 = 1300$ Number of figures after point $= 2 + 1 = 3$).

Decimals are divided as follows.

Example $3 \cdot 76 \div 0 \cdot 4$

1 Remove the 0's from in front of the number you are dividing by and move the point to the right until it becomes a whole number. $3 \cdot 76 \div 4$
2 Move the decimal point the same number of places to the right in the number you are dividing. $37 \cdot 6 \div 4$
3 Now divide as you were shown on page 42. **9·4**

Example

a $2 \cdot 439 \div 0 \cdot 03 = 243 \cdot 9 \div 3 = \mathbf{81 \cdot 3}$
   (Points moved two places to the right before dividing).

b $23 \div 0 \cdot 06$ First we write 23 as $23 \cdot 0000$
   $23 \div 0 \cdot 06 = 23 \cdot 0000 \div 0 \cdot 06 = 2300 \cdot 00 \div 6 = \mathbf{383 \cdot 33}$
   (Points moved two places to right)

page 46 / unit 7

Exercise 7.3

Do these multiplication sums. For example sum number **2e** is 0.7×2.3

		a 0·6	b 0·7	c 0·4	d 1·1	e 2·3	f 3·5	g 0·73
1	0·5							
2	0·7							
3	0·8							
4	1·3							
5	2·4							
6	0·46							
7	1·34							
8	6·31							
9	0·346							
10	23·4							

Do these division sums.

				a	b	c	d
11	Divide	2·4	by	0·2	0·3	0·4	0·06
12		3·6	by	0·3	0·4	0·6	0·03
13		1·44	by	0·4	0·04	0·8	0·03
14		72	by	0·3	0·02	0·004	0·006
15		1·44	by	1·2	2·4	0·24	0·12
16		0·132	by	1·1	0·12	1·2	0·11
17		19·5	by	1·5	0·15	1·3	0·13
18		37	by	1·7	0·16	0·023	2·01
19		25·67	by	2·3	0·45	0·024	2·04
20		894	by	0·34	0·032	1·22	10·5

Exercise 7.4

In each of these examples you have to decide whether to multiply or divide. If a division sum does not come out exactly give the answer to three figures.

1. Find the cost of 0·9 kg of metal if one kg costs £3.
2. How many packets each holding 0·4 kg can be filled with 5·2 kg?
3. How many cakes costing £0·03 each can be bought for £1·14?
4. If a metre of wire costs £0·07, find the cost of 2·3 m.
5. How many pieces of fencing each 0·8 m long will be needed to fill a 16 m gap?
6. 9·8 tonnes of soil have to be moved using a van which can carry 0·7 tonnes. How many journeys will be needed?
7. A litre of mercury weighs 13·6 kg. What is the weight of 0·4 litres?
8. How many bottles of lemonade each costing £0·09 can be bought for £1·80?
9. Find the cost of 2·5 kg of sugar if one kg costs £0·07.
10. How many kg of sugar can be bought for £1·40 if one kg costs £0·07?
11. Find the cost of 5·5 kg of butter if one kg costs £0·34?
12. How many kg of butter can be bought for £5·40 if each kg costs £0·36?
13. How many truck loads will be required to move 27 tonnes of earth, if a truck carries 1·5 tonnes?
14. Find the cost of 6·6 m of wood at £0·74 per metre.
15. Find the cost of 0·75 m of wood at £0·56 per metre.
16. How many bags of sand each holding 0·8 cubic metres can be filled with 13 cubic metres?
17. Find the cost of covering 16·7 square metres of floor at £1·23 per square metre.
18. How many bottles of wine each costing £0·76 can be bought for £27·50?
19. What will be the cost of 5·6 litre of wine if one litre costs £1·24?
20. Assuming that an average person weighs 0·12 tonnes, how many people can travel in a lift which can carry 3·5 tonnes?

Multiplying and dividing numbers by 10 and 100

Multiply 1·345 by 10 and 100. You should get 13·45 and 134·5.

You will notice that the numbers in the answer are the same, but the point has moved to the right.

It has moved one place when we multiplied by 10.

It has moved two places when we multiplied by 100.

If you do some more examples like this you will soon discover the following rules.

to multiply a number by 10 move the point one place to the right

to multiply a number by 100 move the point two places to the right

to multiply a number by 1000 move the point three places to the right

Example

a $100 \times 2·35 = 235·00$.   This is usually written 235

b $1000 \times 1·6 = 1000 \times 1·600 = 1600·00$.   This is usually written 1600

Notice how 1·6 is written as 1·600 before we start to move the point.

c $100 \times 32 = 100 \times 32·00 = 3200·00 = 3200$.

Divide 23·4 by 10 and 100. You should get 2·34 and 0·234.

You will notice that the numbers in the answer are the same, but the point has moved to the left.

It moved one place when we divided by 10.

It moved two places when we divided by 100.

If you do some more examples like this you will soon discover the following rules.

unit 7/page 49

to divide a number by 10
move the point
one place to the left

to divide a number by 100
move the point
two places to the left

to divide a number by 1000
move the point
three places to the left

Example

a $23·5 \div 10 = 2·35$.
b $45·6 \div 100 = ·456 = 0·456$.
c $3·4 \div 1000 = 0003·4 \div 1000 = 0·0034$. Notice how extra 0's are placed in front of the 3 before we move the point along.

Exercise 7.5

Do these multiplication and division sums. For example **2c** is $3·45 \times 1000$.

	a ×10	b ×100	c ×1000	d ÷10	e ÷100	f ÷1000
1 2·7						
2 3·45						
3 8						
4 0·7						
5 25·7						
6 78						
7 0·73						
8 45·67						
9 1·785						
10 0·834						

Cancelling fractions

Look at the three fractions $\frac{2}{4}$, $\frac{5}{10}$, $\frac{6}{12}$. Each one is equal to

Dividing the top and bottom of the first one by 2 we get $\frac{1}{2}$.

Dividing the top and bottom of the next by 5 we get $\frac{1}{2}$.

Dividing the top and bottom of the last by 6 we get $\frac{1}{2}$.

Look at the three fractions $\frac{6}{8}$, $\frac{15}{20}$, $\frac{12}{16}$.

Each one is obviously equal to $\frac{3}{4}$.

Dividing the top and bottom of the first by 2 we get $\frac{3}{4}$.

Dividing the top and bottom of the second by 5 we get $\frac{3}{4}$.

Dividing the top and bottom of the third by 4 we get $\frac{3}{4}$.

It can be shown that if we can divide the top and bottom of a fraction by the *same* number, then the fraction still has the same value, although it may look different. This is called **cancelling** or **simplifying** the fraction.

Example

Simplify these fractions.   a $\frac{14}{16}$ b $\frac{15}{25}$ c $\frac{18}{24}$.

a 2 will divide into top and bottom.

You write it like this. $\frac{\cancel{14}^{7}}{\cancel{16}_{8}} = \frac{7}{8}$.

b 5 will divide into top and bottom.

You write it like this. $\frac{\cancel{15}^{3}}{\cancel{25}_{5}} = \frac{3}{5}$.

c 6 will divide into top and bottom.

You write it like this. $\frac{\cancel{18}^{3}}{\cancel{24}_{4}} = \frac{3}{4}$.

You could also divide 2 into top and bottom, and then 3 into top and bottom writing it like this.

$$\frac{\cancel{\cancel{18}^{9}}^{3}}{\cancel{\cancel{24}_{12}}_{4}} = \frac{3}{4}.$$

unit 7 / page 51

Example

> There are 12 boys in a class and 6 girls. (a) What fraction of the class are boys? (b) What fraction are girls? Simplify each answer. There are 18 in the class so we can say:
>
> a Fraction of boys $= \dfrac{\cancel{12}^{2}}{\cancel{18}_{3}} = \dfrac{2}{3}$.
>
> b Fraction of girls $= \dfrac{\cancel{6}^{1}}{\cancel{18}_{3}} = \dfrac{1}{3}$.

Exercise 7.6

Simplify these fractions:

	a	b	c	d	e	f	g	h	i
1	$\dfrac{2}{4}$	$\dfrac{3}{12}$	$\dfrac{5}{10}$	$\dfrac{15}{20}$	$\dfrac{2}{6}$	$\dfrac{6}{9}$	$\dfrac{4}{12}$	$\dfrac{12}{16}$	$\dfrac{5}{20}$
2	$\dfrac{3}{6}$	$\dfrac{4}{16}$	$\dfrac{3}{9}$	$\dfrac{10}{15}$	$\dfrac{6}{18}$	$\dfrac{6}{24}$	$\dfrac{2}{8}$	$\dfrac{6}{8}$	$\dfrac{5}{15}$
3	$\dfrac{4}{8}$	$\dfrac{9}{12}$	$\dfrac{6}{12}$	$\dfrac{18}{24}$	$\dfrac{6}{9}$	$\dfrac{8}{12}$	$\dfrac{12}{18}$	$\dfrac{21}{28}$	$\dfrac{14}{21}$

In each of the following examples write down what fraction of the class are boys, and what fraction are girls. **Simplify your answers.**

4 3 boys, 9 girls.   5 6 boys, 2 girls.
6 2 boys, 4 girls.   7 12 boys, 4 girls.
8 5 boys, 10 girls. 9 18 boys, 6 girls.

10 A car park contains 12 red cars, 8 blue cars, and 4 green cars.
a What fraction of the cars are red?
b What fraction of the cars are blue?
c What fraction of the cars are green?
Simplify your answers.

11 A car park contains 36 cars. 6 of them are red, 12 are blue, 15 are green, and 3 are black.
a What fraction are red?
b What fraction are blue?
c What fraction are green?
d What fraction are black?
Simplify your answers.

12 48 pupils enter for an examination. 32 pass.
a What fraction pass?
b What fraction fail?

13 64 pupils enter for an examination. 12 fail.
a What fraction pass?
b What fraction fail?

14 35 plants are planted in a garden. 10 of the plants failed to grow.
a What fraction of the plants failed to grow?
b What fraction of the plants did grow?

Changing fractions to decimals

A man needs $\frac{4}{5}$ metres of wood. How would he ask for this in a shop since they only sell wood by the metre, or tenths and hundredths of a metre?

One way is to change $\frac{4}{5}$ into a decimal. It can be shown that it is equal to 0·8. The man has to ask for 0·8 or $\frac{8}{10}$ of a metre.

What we need is a rule for changing any fraction into a decimal. It can be shown that the rule is:

to change a fraction to a decimal divide the top by the bottom

Example

Change these fractions to decimals.

a $\frac{4}{5}$ Write 4 as 4·0 and then divide.

$$\frac{4}{5} = 0·8$$

```
  0·8
5)4·0
```

b $\frac{5}{8}$ Write 5 as 5·000 and then divide.

$$\frac{5}{8} = 0·625$$

```
   0·625
8)5·000
   48
   ‾‾
    20
    16
    ‾‾
    40
```

c $\frac{13}{7}$ Write 13 as 13·000 and then divide.

$$\frac{13}{7} = 1·85$$

```
   1·85
7)13·000
   7
   ‾‾
   6·0
   5·6
   ‾‾
    40
    35
    ‾‾
     5
```

This example does not come out exactly, and so the division is stopped after three figures. This is accurate enough for most purposes.

Exercise 7.7

Change these fractions to decimals. If they do not come out exactly give the answer to three figures.

	a	b	c	d	e	f	g	h	i	j
1	$\frac{1}{2}$	$\frac{1}{5}$	$\frac{1}{4}$	$\frac{2}{5}$	$\frac{3}{4}$	$\frac{1}{3}$	$\frac{3}{5}$	$\frac{4}{5}$	$\frac{2}{3}$	$\frac{1}{6}$
2	$\frac{3}{7}$	$\frac{2}{7}$	$\frac{2}{6}$	$\frac{1}{7}$	$\frac{3}{6}$	$\frac{5}{7}$	$\frac{4}{6}$	$\frac{4}{7}$	$\frac{5}{6}$	$\frac{6}{7}$
3	$\frac{3}{2}$	$\frac{7}{9}$	$\frac{9}{4}$	$\frac{13}{9}$	$\frac{11}{5}$	$\frac{7}{4}$	$\frac{9}{5}$	$\frac{6}{11}$	$\frac{3}{14}$	$\frac{5}{11}$

unit 7/page 53

4 $\frac{13}{27}$ $\frac{7}{3}$ $\frac{27}{35}$ $\frac{12}{5}$ $\frac{81}{87}$ $\frac{19}{32}$ $\frac{33}{64}$ $\frac{9}{14}$ $\frac{81}{89}$ $\frac{42}{91}$

5 $\frac{3}{65}$ $\frac{7}{42}$ $\frac{22}{37}$ $\frac{43}{87}$ $\frac{83}{126}$ $\frac{72}{425}$ $\frac{96}{327}$ $\frac{3}{896}$ $\frac{241}{462}$ $\frac{900}{874}$

Sometimes it may be possible to simplify the fraction by cancelling before you divide.

Example

$$\frac{\cancel{15}^3}{\cancel{20}_4} = \frac{3}{4} = 0{\cdot}75.$$

Do these the same way.

	a	b	c	d	e	f	g	h	i	j
6	$\frac{4}{6}$	$\frac{4}{8}$	$\frac{12}{20}$	$\frac{21}{12}$	$\frac{8}{10}$	$\frac{2}{10}$	$\frac{2}{8}$	$\frac{4}{12}$	$\frac{3}{6}$	$\frac{32}{12}$
7	$\frac{6}{15}$	$\frac{12}{16}$	$\frac{20}{35}$	$\frac{42}{30}$	$\frac{72}{66}$	$\frac{42}{77}$	$\frac{10}{35}$	$\frac{40}{45}$	$\frac{54}{24}$	$\frac{25}{45}$

Finding fractions of numbers

There are 36 pupils in a class. The teacher decides to take $\frac{3}{4}$ of them out one day. How many does he take?

He has to work out $\frac{3}{4}$ of 36 or $\frac{3}{4} \times 36$.

$$\frac{3}{4} \times 36 = \frac{3}{4} \times \frac{36}{1} = \frac{3 \times 36}{4 \times 1} = \frac{108}{4} = 108 \div 4 = 27.$$

Sometimes examples like this can also be done using cancelling.

$$\frac{3}{4} \times 36 = \frac{3}{{}_1\cancel{4}} \times \frac{\cancel{36}^9}{1} = \frac{27}{1} = 27 \quad \text{(here 4 is cancelled)}.$$

Example

Find $\frac{5}{8}$ of 64

$$\frac{5}{{}_1\cancel{8}} \times \frac{\cancel{64}^8}{1} = \frac{40}{1} = 40 \quad \text{(here 8 is cancelled)}.$$

Example

A man decides he can save $\frac{2}{7}$ of what he earns. How much will he save out of £22·50?

$$\frac{2}{7} \times 22{\cdot}50 = \frac{2}{7} \times \frac{22{\cdot}50}{1} = \frac{45{\cdot}00}{7} = 45{\cdot}00 \div 7 = £6{\cdot}42.$$

This example does not come out exactly, so it is worked out to give the two figures needed for the pence.

Exercise 7.8

		a	b	c	d	e
1	Find $\frac{3}{4}$ of	12	16	8	24	28
2	Find $\frac{2}{5}$ of	10	15	25	40	60
3	Find $\frac{1}{3}$ of	12	15	24	30	66
4	Find $\frac{5}{7}$ of	14	21	25	30	46
5	Find $\frac{7}{9}$ of	26	56	98	29	84
6	Find $\frac{17}{23}$ of	27	65	92	63	184
7	Find $\frac{5}{8}$ of	£1·20	£1·52	£4·38	£0·83	
8	Find $\frac{2}{7}$ of	£3·15	£4·69	£0·68	£34·89	
9	Find $\frac{11}{23}$ of	£23·87	£47·98	£0·98	£25	

10 There are 24 boys and girls in a class. The teacher decided to let $\frac{1}{4}$ of the class go to the local library every Thursday. How many go each time?

11 There are 48 boys and girls in a small school. Two-thirds of them go swimming. How many go swimming?

12 A man decides to save $\frac{1}{5}$ of what he earns. How much does he save if he earns
a £22·50 b £23·25 c £42 d £13·49 e £45·84?

13 After a shopkeeper has fixed the price of what he sells he must add Value Added Tax of one tenth onto his price. One way of doing this is to multiply his price by $\frac{11}{10}$ to get the selling price. Find the selling price of the following items (the shopkeeper's price is given)
a £2·50 b £8·60 c £0·80 d £3·45 e £24·80?

14 Two men who own a shop agree to share the profits so that one of them gets $\frac{2}{7}$ of the profits, and the other gets $\frac{5}{7}$ of the profits. How should they share out profits of:
a £28 b £15·40 c £44·10 d £26 e £24·56?

unit 8 Fixing Positions with Numbers

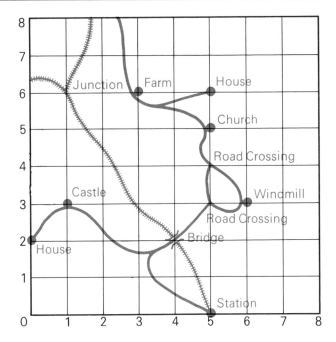

If you look at the map you will see that there is only one castle, but there are two houses. If we want to say which house we are talking about we can do it by giving two numbers. If we start at the bottom left-hand corner of the map you will see that we need to go 5 squares along, and then 6 squares up to get to one of the houses. We can say that the house is at (5, 6). You should notice that the **first** number gives the number of squares to the right, and the **second** number gives the number of squares up from the base line.

To get to the windmill we need to start at 0 and go 6 squares to the right, and then 3 squares up. We can say that the windmill is at (6, 3).

Copy and complete these:

The castle is at (,)
The farm is at (,)
The bridge is at (,)
The railway junction is at (,)
The other house is at (,)
The station is at (,)
The church is at (,)
The two road crossings are at (,) and at (,)

Exercise 8.1

1 Using squared paper mark the numbers 0 to 8 along the bottom, and 0 to 8 up the side. Mark the position of each of the following with a dot and write the name beside it.

Station (8, 5) Windmill (2, 7) Church (3, 5) Farm (0, 5)
Station (5, 0) House (2, 1) House (6, 7) Castle (0, 3)

Draw a straight road from the church to the farm.
Draw a straight road from the church to the first house.
Draw a straight road from the church to the second house.
Draw a straight railway line between the stations.

2 Using another piece of squared paper mark the numbers 1 to 10 along the bottom and 1 to 10 up the side. Mark in these points:

A (2, 6) B (2, 9) C (4, 7) D (6, 5) E (5, 6)
F (5, 2) G (7, 2) H (2, 3) I (1, 6) J (1, 9)
K (0, 9) L (3, 2) M (5, 0) N (7, 0) O (3, 0)
P (1, 3) Q (0, 3) R (0, 6)

When we say 'Mark the point A (2, 6)' this means that we make a dot at the point (2, 6) and write the letter A by it.

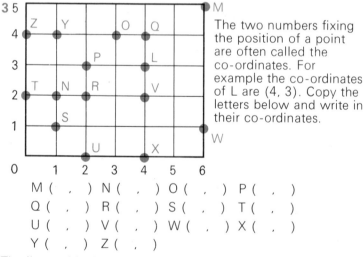

The two numbers fixing the position of a point are often called the co-ordinates. For example the co-ordinates of L are (4, 3). Copy the letters below and write in their co-ordinates.

M (,) N (,) O (,) P (,)
Q (,) R (,) S (,) T (,)
U (,) V (,) W (,) X (,)
Y (,) Z (,)

The lines with the numbers on are called the axes. If the axes are placed in the middle of the map instead of along the edges, then negative numbers have to be used to give the positions of some of the points.

Looking at the map on p. 57:

Starting at 0 we need to go 2 squares *to the right* to get to the windmill, but instead of going up to it we have to go 1 square *down*. If you remember the work that we did on the number line in unit 6, you will see that we can write 1 down as ⁻1. We can say that the co-ordinates of the windmill are (2, ⁻1).

unit 8/page 57

To get to the junction we have to go 3 squares *to the left*. We can write this as ⁻3. We also go 2 squares up, so we can say that the co-ordinates of the junction are (⁻3, 2).

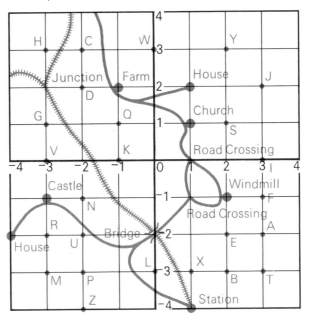

To get to the castle we have to go 3 squares *to the left*, and then 1 square *down*. We can say that the co-ordinates of the castle are (⁻3, ⁻1).

If we have to go *to the left* instead of forwards the first co-ordinate is *negative*. If we have to go *down* instead of up the second co-ordinate is *negative*.

Copy and complete these:

The church is at (,)   The station is at (,)
The bridge is at (,)   The farm is at (,)
The houses are at (,) and at (,)
The road crossings are at (,) and at (,)

Exercise 8.2

1 Points on the map have been marked with the letters from A to Z. Write down the co-ordinates of each of these points.

2 On a piece of squared paper draw the axes in the middle of the sheet, and on both axes mark the numbers from ⁻5 to 5. Mark in these points.

a (1, 3) b (3, 2) c (⁻3, 2) d (⁻4, 3) e (1, ⁻2)
f (3, ⁻4) g (⁻4, ⁻2) h (⁻2, ⁻4) i (0, 3) j (0, 1)
k (3, 0) l (0, 5) m (⁻4, 0) n (⁻3, 0) o (4, 3)
p (⁻1, ⁻3) q (⁻1, 3) r (0, 4) s (0, ⁻4)
t (⁻1, ⁻5) u (⁻3, ⁻1) v (⁻5, ⁻1) w (5, ⁻5)
x (⁻3, ⁻5) y (3, ⁻2) z (5, ⁻1)

unit 9 Area and Volume

What is area?

Look at these three shapes. Which is the largest?

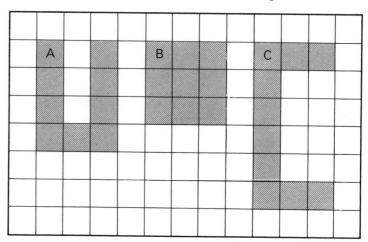

If you count the squares you will find that C is the largest with 10 squares and A and B both have 9 squares. You should notice that we are measuring the amount of space inside each shape. This is called the **area**.

The area of C = 10 squares.

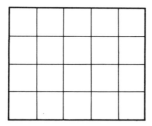

This shape is called a rectangle. Count the number of squares inside to find its area. You will find that its area is 20 squares. The length of the rectangle is 5, and the width is 4, and you can see that instead of counting the squares we could have said

Area = 5 × 4 = 20 squares.

The area of any rectangle can be found in the same way and we can write:

area of rectangle = length × width

unit 9/page 59

Example

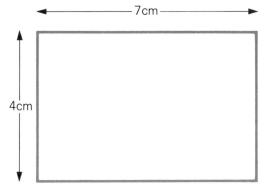

Find the area of this rectangle

Area = 4 × 7 = 28 square cm.

Notice that the answer is given in square centimetres. This means that if we cut up the rectangle into squares of side 1 cm we would get 28 squares. If a rectangle is measured in metres its area will be measured in square metres.

Exercise 9.1

1 Count the number of squares in the three shapes; E, I, and H.

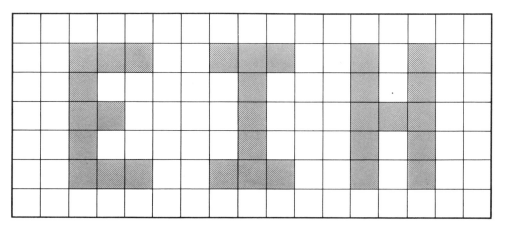

a Which has the greatest area?
b Which has the least area?

2. Find the areas of the following rectangles:

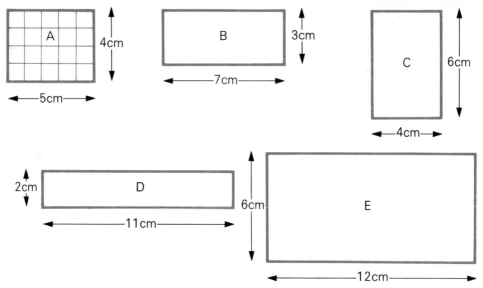

3. Work out the areas of rectangles with the following measurements:

   a 3 cm by 5 cm   b 4 cm by 4 cm   c 6 cm by 5 cm
   d 9 cm by 7 cm   e 10 cm by 12 cm
   f 15 m by 20 m   g 4 cm by 2·5 cm
   h 5 m by 3·5 m   i 1·5 m by 1·5 m
   j 1·6 cm by 3·2 cm   k 3·7 m by 2·9 m

4. A man decides to carpet his living room. The room is 4 m by 5 m, and the cost of the carpet is £3 per square metre.

   a What is the area of the living room?
   b What will be the cost of the carpet?

5. A man decides to cover his back yard with concrete. The yard is 2 m by 7 m, and he buys the concrete ready mixed at 50 p per bag. Each bag covers 2 square metres.

   a What is the area of the back yard?
   b How many bags of concrete will he need to buy?
   c How much will the concrete cost?

6. A fence is 12 m long and 2 m high. It is to be painted using tins of paint, each one of which covers 4 square metres. Each tin of paint costs 27p.

   a What is the area of the fence?
   b How many tins of paint will be needed?
   c How much will the paint cost?

7 A woman decides to carpet her living-room. The room is 4 m by 4·5 m and the price of the carpet that she wishes to buy is £3 per square metre.
   a What is the area of the living-room?
   b How much will she have to pay for the carpet?

8 The path in a man's garden is 0·75 m wide and 8 m long. He puts cement on the path using one 50 p bag for each 1·5 square metres.
   a What is the area of the path?
   b How many bags of cement does he use?
   c What is the cost of cementing the path?

Finding the area of a triangle

Look at these pictures which show how you can take two triangles which are the same size and shape, cut one of them up, and join them together to form a rectangle.

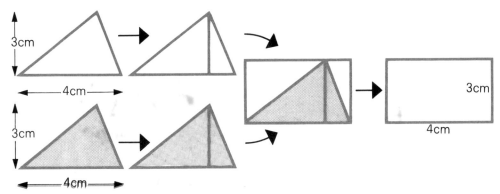

Since the areas do not change, we can say:
   Area of the two triangles = area of rectangle
   = 4 × 3 = 12 square cm.

So we can write:
   Area of one triangle = ½ × area of rectangle
   = ½ × 4 × 3 = ½ × 12 = 6 square cm.

The **base** of the triangle is 4 cm, and the **height** is 3 cm. We can obviously find the area of any triangle the same way so we can write:

area of triangle = ½ × base × height

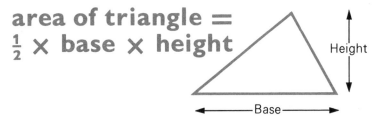

page 62 / unit 9

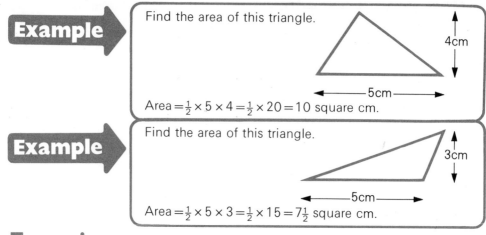

Example — Find the area of this triangle.
Area = ½ × 5 × 4 = ½ × 20 = 10 square cm.

Example — Find the area of this triangle.
Area = ½ × 5 × 3 = ½ × 15 = 7½ square cm.

Exercise 9.2

Find the areas of the following triangles.

1

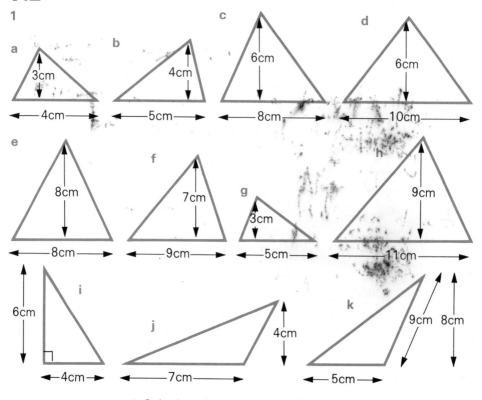

2 Calculate the areas of the following triangles:
  a base 2 cm, height 4 cm   b base 12 cm, height 4 cm
  c base 5 m, height 6 m   d base 8 cm, height 7 cm
  e base 7 m, height 9 m   f base 14 cm, height 12 cm

g base 11 cm, height 17 cm   h base 1·2 cm, height 5 cm   i base 3·2 cm, height 2·7 cm
j base 4·7 cm, height 5·3 cm

3 The drawing shows part of the end of a house. It is to be covered with a special protective paint at 9p per square metre.

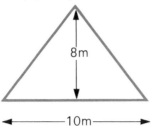

a What is the area to be covered with paint?
b What will be the cost?

Repeat question 3 with the following areas.

4 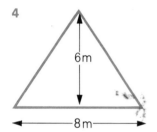   5 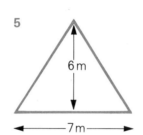   6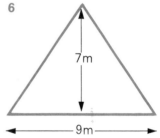

7 a Measure the base of the triangle (from A to B), and the height to the nearest tenth of a cm. Find its area.

b Turn the page round so that BC is now the base. Measure the new base and new height and find the area.

c Turn the page round so that CA is now the base. Measure the new base and new height and find the area.

d Should the answers to a, b, and c be the same? Are yours?

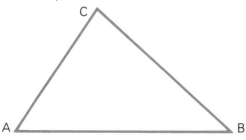

8 The flower bed in Mr. Jones' front garden is shaped as in the diagram. Fertilizer is to be spread on the flower bed at a rate of 250 g per square metre.

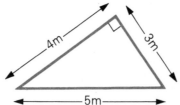

a What is the area of the flower bed?
b What weight of fertilizer will be required?
c What will be the cost of the fertilizer if 1 kg costs 73p?

Finding the areas of other shapes

If all the sides of a shape are straight lines it is often possible to find the area by cutting it up into rectangles and triangles, finding the area of each part, and then adding the areas together. The best way to do this is to draw the shapes again divided into the rectangles and triangles, and to mark in the measurements of each part. Three examples are done below.

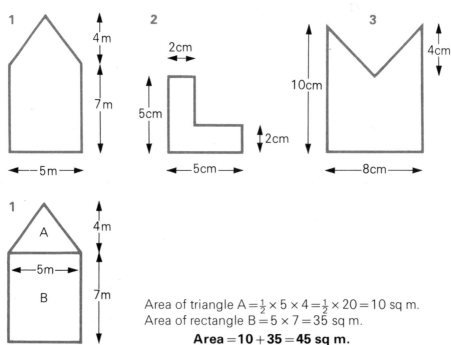

Area of triangle A = $\frac{1}{2}$ × 5 × 4 = $\frac{1}{2}$ × 20 = 10 sq m.
Area of rectangle B = 5 × 7 = 35 sq m.
Area = 10 + 35 = 45 sq m.

2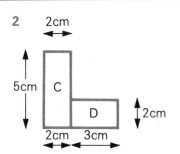

Area of rectangle C = 5 × 2 = 10 sq cm.
Area of rectangle D = 3 × 2 = 6 sq. cm.
Area = 10 + 6 = 16 sq cm.

3

In this case we find the area of the rectangle and **take away** the area of the triangle.

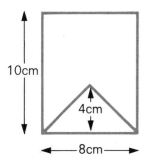

Area of rectangle = 10 × 8 = 80 sq cm.
Area of triangle = $\frac{1}{2}$ × 8 × 4 = $\frac{1}{2}$ × 32 = 16 sq cm.
Area = 80 − 16 = 64 sq cm.

Exercise 9.3

1 Find the areas of the following shapes, by dividing them into rectangles:

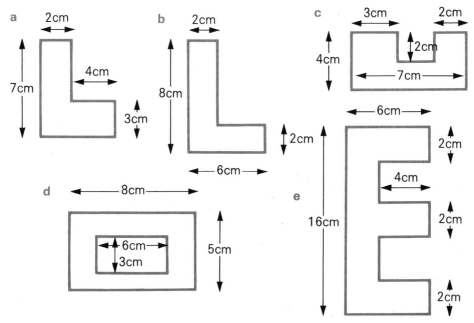

2 A man has a lawn in the shape of a rectangle measuring 9 m by 11 m. He decides to turn the lawn into a garden leaving a strip 1 m wide along the four sides.
  a Draw a plan of the garden with the strips of grass around it.
  b Find the area of the grass strips.

3 A piece of wood measuring 8 m by 2·5 m intended as the front wall of a shed, has two window spaces 0·8 m by 0·6 m and a door space 2 m by 0·6 m cut from it. What is the area of wood left in the wall?

4 Find the areas of the following shapes, by dividing them into rectangles and triangles.

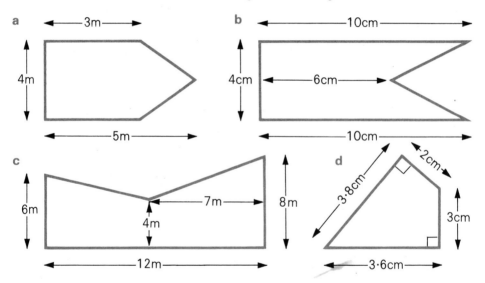

5 A kite 80 cm long and 55 cm wide is cut from a rectangular piece of cloth 1 m by 60 cm.
  a Draw a diagram of the cloth with the kite cut out.
  b Find the area of cloth left.

The volume of a rectangular block

How much space is there in this rectangular block?

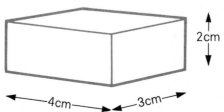

unit 9/page 67

One way of answering this question is to imagine the block cut up into small cubes with sides of 1 cm.

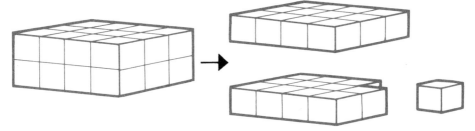

You will see that in this case there are 24 cubes, so we can say:

Volume of block = **24 cubic cm.**

There is no need to make a drawing to do this as we can get the same result by multiplying the length (4 cm), by the width (3 cm), by the height (2 cm). The volume of any block can be found the same way and we can write:

volume of block = length × width × height

Find the volume of this block.

Volume = 5 × 4 × 3 = 5 × 12 = **60 cubic cm.**

You can see by the drawing that this is the answer we get if we cut the block up and count the cubes. There are 20 cubes on the top, 20 in the middle, and 20 on the bottom. A total of 60.

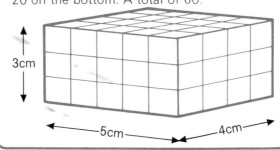

Exercise 9.4

Find the volumes of the following rectangular blocks:

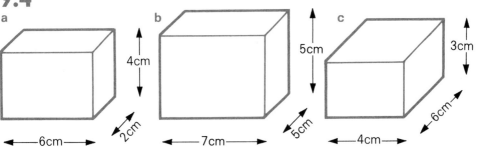

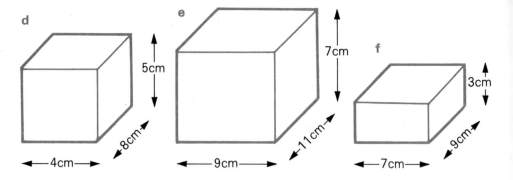

2 Calculate the volumes of rectangular blocks with the following measurements:

	Length	Width	Height
a	6 cm	8 cm	3 cm
b	3·5 cm	4 cm	5 cm
c	2·7 m	3 m	2·5 m
d	19 cm	13 cm	11·3 cm
e	23 cm	19 cm	17 cm

3 In each of the following examples find the volume of the room, and the number of people that can work in the room if each person must have 6 cubic metres of space.

	Length	Width	Height
a	9 m	6 m	2 m
b	10 m	8 m	3 m
c	12 m	8 m	4 m
d	20 m	10 m	3 m
e	17 m	9 m	3 m

4 Wooden play blocks just fill a rectangular box with inside measurements 80 cm by 70 cm by 35 cm. The blocks are cubes 5 cm by 5 cm by 5 cm.
   a What is the volume of the box?
   b What is the volume of a block?
   c How many blocks are there?

5 A brick measures 24 cm by 12 cm by 8 cm. How many bricks would be required to build a wall 12 cm thick, 30 m long and 1·6 m high? Ignore the thickness of the cement.

The surface area of a rectangular block

If we wanted to make a garden shed with a flat roof like the one shown below, how much wood will we need (including enough for the floor)?

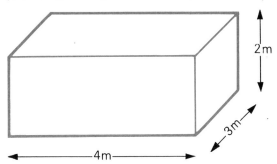

The shed is made up of six rectangles. The roof, the floor, the front, the back, and the two sides. If we find the area of each one and add them together we will find the total surface area of the shed.

Area of roof = 4 × 3 = 12 sq m.
Area of floor = 4 × 3 = 12 sq m.
Area of front = 4 × 2 = 8 sq m.
Area of back = 4 × 2 = 8 sq m.
Area of one side = 3 × 2 = 6 sq m.
Area of second side = 3 × 2 = 6 sq m.
   Surface area of shed = **52 sq m.**

We would have to buy 52 square metres of wood. We would also have to buy some wood to make the frame, and we would need to allow for doors and windows.

Exercise 9.5

1 Find the surface area of these rectangular blocks:

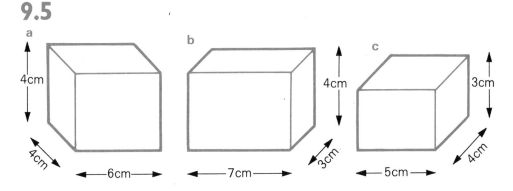

2 Calculate the surface areas of the following rectangular blocks:

	Length	Width	Height
a	6 cm	8 cm	3 cm
b	5 cm	4 cm	5 cm
c	7 m	4 m	5 m
d	19 cm	13 cm	11 cm
e	23 cm	19 cm	17 cm

3 A rectangular oak chest has outside measurements 1·4 m by 0·6 m by 0·4 m.

  a Find the surface area of the chest (i) in square metres and (ii) in square centimetres.

  b How many tins of varnish each containing 60 cc will be needed to varnish the chest, if 1 cc will cover 100 square centimetres?

  c If each tin of varnish costs 12p, what will be the cost of varnishing the chest?

unit 10 Using Letters Instead of Numbers

Areas

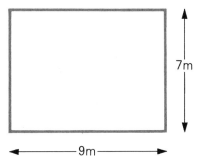

The area of this rectangle is 7 × 9 = 63 square metres.
If you wanted to tell someone how to find the area of a rectangle you can write or say

Area = Length × Width

this can also be written in a shorter form as

$A = LW$.

This is called the **formula** for the area of a rectangle, and you will notice that the multiplication sign has been missed out.

Perimeters

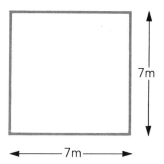

The distance around a shape is called the **perimeter**.
The distance around this square is 4 × 7 cm = 28 cm.
If you wanted to tell someone how to find the distance around a **square** you can write or say

Distance around the square = 4 × side

this can also be written in a shorter form as

$D = 4S$.

This is called the formula for the distance around a square.

Again you will notice that the multiplication sign has been missed out. This is often done when writing down a formula, but it is important to remember that other signs like $+$, $-$, and $\div$ are never missed out.

The distance around the **rectangle** (p. 71) is $7+7+9+9=32$ m.

If you want to tell someone how to find the distance around a rectangle you can write or say

Distance around=twice the width+twice the length

this can be written in a shorter way as

$D = 2W + 2L$.

This is called the formula for the distance around a rectangle.

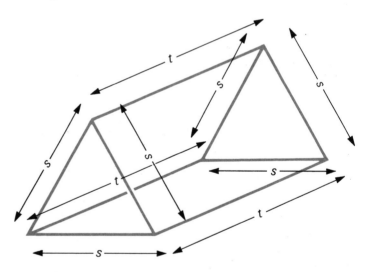

The diagram shows a frame made out of wire.
The length of wire needed $= s+s+s+s+s+s+t+t+t$.
We can write this in a shorter way as

$L = 6s + 3t$.

unit 10/page 73

Exercise 10.1

Write down formulae for the distance around these shapes. Start each answer with $D=$

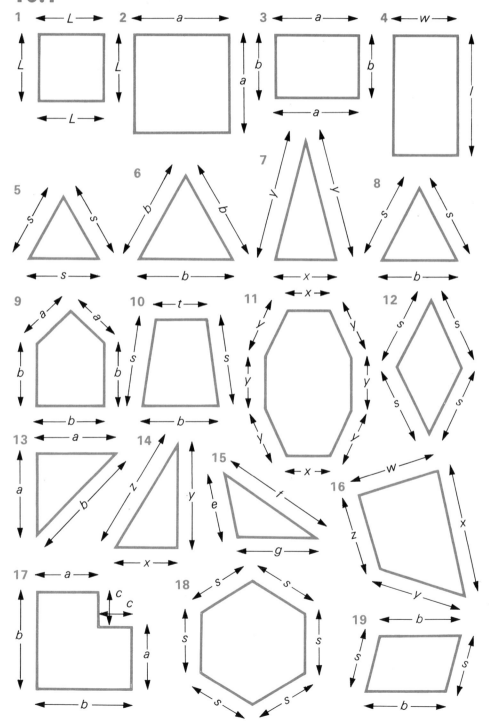

page 74/unit 10

Write down formulae for the length of wire needed for each of these frameworks. Start each answer with $W=$

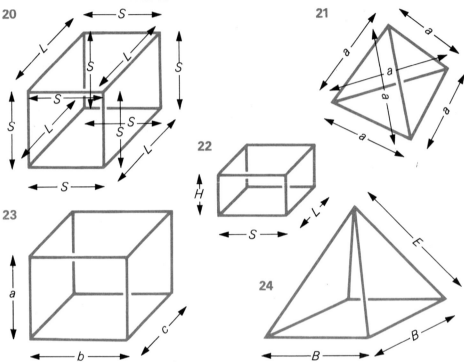

It is possible to have formulae for other things besides lengths, areas, and volumes.

Example

If B stands for the number of boys in a class, G stands for the number of girls in a class, and T stands for the total number of boys and girls in the class; write down a formula starting $T=$
Since we get the total number of pupils by adding the number of boys to the number of girls the answer is

$$T = B+G \quad \text{or} \quad T = G+B \quad \text{both are correct.}$$

Example

If B, G, and T stand for the same things as in the previous examples, write down formulae starting with:
a $B=$ b $G=$

a Since we get the number of boys by taking the number of girls from the total number in the class the answer is

$$B = T - G \quad \text{(Why is it incorrect to say } B = G - T\text{?)}$$

b Since we get the number of girls by taking the number of boys from the total number in the class the answer is

$$G = T - B \quad \text{(Why is } G = B - T \text{ incorrect?)}$$

unit 10/page 75

Example

A boy buys P pencils which cost him C pence each. The total cost is T pence. Write down formulae starting with **a** $T=$ **b** $C=$

a The total cost is got by multiplying the cost of one pencil by the number of pencils so we can write
$$T = P \times C \quad \text{or} \quad T = C \times P$$
we can miss out the multiplication signs and write
$$T = PC \quad \text{or} \quad T = CP$$

b To find the cost of one pencil we need to divide the cost of all the pencils by the number of pencils. The answer is
$$C = T \div P \quad \text{or} \quad C = \frac{T}{P} \quad \text{(Why is } C = P \div T \text{ incorrect?)}$$

Exercise 10.2

In the first ten questions there are four answers to each question. Two of them are correct, and two of them are wrong. Write down the two *correct* ones.

1 N is the number of pupils in a school. G is the number of girls. B is the number of boys.
$G = B - N \quad N = B + G \quad B = G - N \quad B = N - G$

2 M stands for the number of men on a coach. W stands for the number of women on the coach. T stands for the total number of men and women on the coach.
$M = T - W \quad M = W - T \quad T = M + W \quad T = M - W$

3 D stands for the number of desks in a room. C stands for the number of chairs in a room. N stands for the number of desks and chairs in the room.
$N = D + C \quad C = N + D \quad N = D - C \quad N = C + D$

4 A box weighs B grams. It contains W grams of washing powder. The weight of the box and the washing powder is T grams.
$W = B + T \quad B = T - W \quad T = B + W \quad B = W + T$

5 A car costs C pounds. A caravan costs X pounds. The total cost is T pounds.
$T = C + X \quad T = C - X \quad X = C - T \quad X = T - C$

6 A girl buys P pens which cost Y pence each. The total cost is T pence.
$T = PY \quad T = YP \quad Y = TP \quad P = TY$

7 A train has N coaches. Each coach carries P people. T is the total number of people on the train.
$T = NP \quad T = N + P \quad N = T \div P \quad N = P \div T$

8 There are B boxes of washing powder in a shop. Each box weighs G grams. The total weight of all the boxes is W.
$W = BG \quad B = W \div G \quad B = G \div W \quad W = B + G$

9 There are R reels of cotton. Each reel contains C cm of cotton. The total length of all the cotton on all the reels is L cm.
   $L = RC \quad R = L \div C \quad R = C + L \quad R = C \div L$

10 There are B boys in a class. The total of their exam marks is T. The average mark is A.
   $T = A + B \quad T = AB \quad A = T \div B \quad A = B \div T$

11 A box weighing B gm is filled with S gm of soap powder. The total weight of the box and soap powder is T gm.
   a Write down for formula starting $T =$
   b Write down the formula starting $S =$
   c Write down the formula starting $B =$

12 A man buys a car costing £C and a boat costing £B. If the total cost is £T
   a Write down the formula starting $T =$
   b Write down the formula starting $C =$
   c Write down the formula starting $B =$

13 A case contains B boxes, and each box contains M matches. If the total number of matches in the case is T.
   a Write down the formula starting $T =$
   b Write down the formula starting $B =$
   c Write down the formula starting $M =$

14 A reel of cotton has N turns of cotton on it. The length of one turn is L cm. The total length of cotton on the reel is R cm.
   a Write down the formula starting $R =$
   b Write down the formula starting $L =$
   c Write down the formula starting $N =$

Example

A man spends £A, then £B, then £C. How much does he spend? We get the answer by adding up the three amounts that he spends.
$A + B + C$ (Other answers such as $B + C + A$ are also correct)

15 A woman spends £X, then £Y, then £Z, and then £W. How much does she spend?

16 A man gets £P out of the bank, then he spends £Y. How much has he left?

17 A man gets £P out of the bank. He spends £X, and then £Y. How much has he left?

18 Find the total cost of P pencils at X pence each, and R rulers at Y pence each.

19 A man works for h hours at p pence per hour.
   a How much does he earn (answer in pence)?
   b How much does he earn (answer in £s)?

20 A man has p pence. He buys b loaves of bread at x pence each, and c cakes at y pence each.
   a How much has he spent?
   b How much has he left?

Making formulae simpler

In the work on formulae you saw that we could write $4a$ instead of $a+a+a+a$. This is often called simplification (making simpler). Other examples are:

$L \times W = LW$
$s+s+s = 3s$
$3 \times m = 3m$
$r \times p = rp$
$L+L+S+S+w+w = 2L+2S+2w$
$4x+7y-2x = 2x+7y$
$n \times p = np$
$W+W+L+L = 2W+2L$
$6b-2b = 4b$

Some formulae cannot be simplified any more. Examples of this are.

$AB \qquad X+Y \qquad 3b-5c \qquad pq \qquad 8d \qquad x+y+z$

When you are simplifying a formula you may only miss out the multiplication signs. **You must not miss out $+$, $-$, or $\div$.**

Here are some other examples of simplification.

Example

$2 \times 6a = 12a \qquad A \times B \times C = ABC \qquad 3M \times N = 3MN$

$12s \div 3 = 4s \qquad 6d \times 2e = 12de \qquad \dfrac{8a}{4} = 2a$

$\dfrac{12ab}{6} = 2ab$

The last two examples are simplified by cancelling. Here are two more examples simplified by cancelling.

Example

$\dfrac{20pq}{5p}$ if we cancel out 5 into the top and bottom and p into the top and bottom we get $4q$.

$\dfrac{12ab}{16ac}$ we can cancel 4 and a giving $\dfrac{3b}{4c}$.

$\dfrac{16ab}{15pq}$ cannot be simplified as there is nothing that will cancel into the top and the bottom.

Exercise 10.3

Simplify the following:

	a	b	c
1	$x+x+x+x$	$x+x+y+y+y$	$x+x+y+y$
2	$a+b+a$	$a+b+b+b+a$	$2a+4a$
3	$A \times B$	$L \times W$	$P \times Q$
4	$a+a+a+a$	$4a-2a$	$6b-2b$
5	$6b-6b$	$5b+6b$	$6c-5c$
6	$5a+2b+4a+7b$	$4m+7n-m$	$5x+7y-2x-3y$
7	$2 \times 4a$	$3 \times 7a$	$6 \times ab$
8	$5a \times 3b$	$8g \times 5f$	$9q \times 2p$
9	$6a \div 2$	$12b \div 3$	$8q \div 2$
10	$\dfrac{8a}{2}$	$\dfrac{16b}{4}$	$\dfrac{15q}{3}$
11	$\dfrac{10a}{a}$	$\dfrac{16b}{b}$	$\dfrac{18q}{q}$
12	$\dfrac{10ab}{5}$	$\dfrac{12pq}{p}$	$\dfrac{14ef}{7e}$
13	$\dfrac{20ab}{4}$	$\dfrac{16pq}{q}$	$\dfrac{15ef}{3f}$

d

1. $y+y+y+y+y$
2. $5a+6a+7a$
3. $4 \times B$
4. $6b-5b$
5. $4a+6a-5a$
6. $a+2a+3a+7b-6a$
7. $a \times b \times c$
8. $2p \times 6q$
9. $10p \div 5$
10. $\dfrac{20p}{4}$
11. $\dfrac{6p}{p}$
12. $\dfrac{6gh}{2gh}$
13. $\dfrac{8gh}{4gh}$

Simplify the following where this is possible. Where it is not possible say so.

	a	b	c	d
14	$3x+4x$	$x+2y$	$x+x+y$	$x+v+w$
15	$3a+4b-2a$	$6a-5b$	$d+b-d$	$a-b+c$
16	$2a \times 3b$	$3 \times 4a$	$5e \times 2$	$4r \times 2f$
17	$\dfrac{4ab}{2ab}$	$\dfrac{5bc}{7ef}$	$\dfrac{7b}{14c}$	$\dfrac{9bc}{4ac}$

unit 11 Using Formulae

In Unit 10 you learnt how to write down simple formulae; in this Unit you will be learning how to use formulae.

Before a formula can be used we need to know what the numbers stand for.

Example

Use the formula $V=LWH$ to find the volume of this block, where V stands for volume, L for length, W for width and H for height.

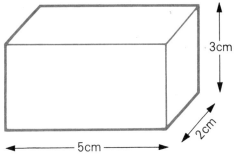

If you look at the block you will see that $L=5$, $W=2$, $H=3$; so we can say
$$V = 5 \times 2 \times 3 = 10 \times 3 = 30 \text{ cubic cm}$$

To give you practice in using formulae you should read the next example, and then try Exercise 11.1.

Example

If $a=7$, $b=5$, $c=2$ find the value of:

1 $3a$ 2 $a+b$ 3 $4a-5c$ 4 $\dfrac{a+b}{4+c}$ 5 $\dfrac{bc}{a}$

1 $3a = 3 \times 7 = 21$

2 $a+b = 7+5 = 12$

3 $4a-5c = 4 \times 7 - 5 \times 2 = 28 - 10 = 18$

4 $\dfrac{a+b}{4+c} = \dfrac{7+5}{4+2} = \dfrac{12}{6} = 2$

5 $\dfrac{bc}{a} = \dfrac{5 \times 2}{7} = \dfrac{10}{7} = 1 \cdot 42$

Exercise 11.1

If $a=8$, $b=5$, $c=3$, $d=1$, $e=0$ find the value of:

	a	b	c	d	e
1	$4a$	$5b$	$7d$	$6e$	$9a$
2	$a+b$	$c+d$	$d+e$	$a+e$	$a+d$
3	$b+5$	$7+a$	$6+e$	$d+11$	$b+9$
4	$a-b$	$c-d$	$b-a$	$c-e$	$a-c$
5	$a-4$	$d-6$	$e-17$	$c-3$	$a-1$
6	$2a+5b$	$4c+3d$	$2b+3c$	$2e+4d$	$3a+b$

7	$2c-b$	$4c-3d$	$2b-3c$	$2e-4d$	$3b-a$
8	ab	bc	ac	bd	be
9	$3ab$	$4bc$	$6ae$	$7cd$	$5ac$
10	$\dfrac{b+c}{2}$	$\dfrac{a+8}{b-c}$	$\dfrac{b-d}{b-c}$	$\dfrac{2a+4}{b}$	$\dfrac{3a+d}{c+2d}$
11	$\dfrac{a+c}{b}$	$\dfrac{a+b}{c}$	$\dfrac{2b+c}{a}$	$\dfrac{a+b+c}{c}$	$\dfrac{2a+b-c}{c+4}$

Example

The area (A) of this shape is given approximately by the formula $A=3xy$. Find the area when $x=4m$ and $y=5m$.

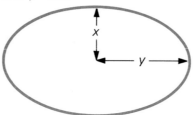

$A = 3 \times 4 \times 5 = 3 \times 20 = 60$ sq. m.

Example

The weight of a bar of iron (W) with length L cm, width W cm, and height H cm is given by the formula $w=8LHW$ grammes. Find the weight of the bar shown.

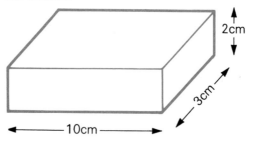

$w = 8 \times 10 \times 3 \times 2 = 80 \times 6 = 480$ grammes.

Example

One way of deciding how long it will take to cook a joint of meat which weighs w kilogrammes is by using the formula $t=20w+30$. This will give the time in minutes. How long will it take to cook a 5-kg joint?

We know that $w=5$ kg, so we can write

$t = 20 \times 5 + 30 = 100 + 30 = 130$ minutes.

unit 11/page 81

Example

The formula for finding the area (A) of the end of a house shaped like the one in the diagram is $A = \frac{1}{2}b(h+H)$. Find the area of the end of a house where $b = 10m$, $h = 6m$, and $H = 8m$.
The brackets mean that the part inside is worked out first.

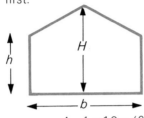

$A = \frac{1}{2} \times 10 \times (6+8) = \frac{1}{2} \times 10 \times 14 = \frac{1}{2} \times 140 = 70$ sq m.

Exercise 11.2

Use the formulae given in the examples above to do the first four examples in this exercise.

1 Find the areas of these shapes.

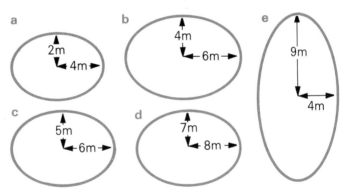

2 Find the weights of these iron bars.

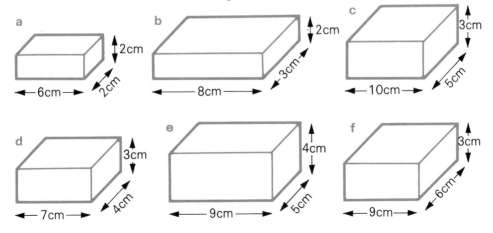

3 How long will it take to cook joints with the following weights?
   a 2 kg   b 1 kg   c 3 kg   d 4 kg   e 6 kg
   f 7 kg   g 8 kg   h 14 kg i 15 kg j 25 kg

4 Find the areas of the ends of these houses.

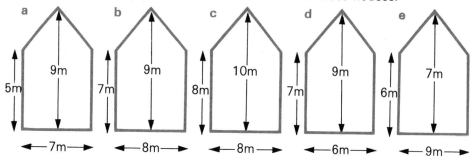

5 The cost (£C) of covering the floor of a room with a certain type of carpet is given by the formula $C = 5LW$, where L and W are the length and the width of the room measured in metres. Find the cost of carpeting rooms with the following measurements.
   a 2 m by 3 m   b 3 m by 4m   c 4 m by 6 m
   d 5 m by 7 m   e 9 m by 8 m

6 If a car travels at a speed of 60 km per hour, the distance (D) it goes in h hours is given by the formula $D = 60h$ kilometres. How far will it travel in the following times?
   a 2 hours   b 3 hours   c 5 hours   d 9 hours
   e 12 hours

7 The cost (£C) of supplying and fixing a certain kind of carpet in a room L metres long and W metres wide is given by $C = 7 + 3LW$. Find the cost of carpeting rooms with the following measurements.
   a 3 m by 4m   b 4 m by 5 m   c 5 m by 5 m
   d 5 m by 7 m   e 8 m by 9 m

8 The area of the walls of a room with a length of L metres, a width of W metres, and a height of H metres is given by $A = 2H(L + W)$ square metres. Find the areas of the walls of the rooms with the following measurements.

	a	b	c	d	e	f
Length	4 m	5 m	7 m	8 m	9 m	10 m
Width	2 m	3 m	4 m	5 m	6 m	8 m
Height	2 m	2 m	3 m	4 m	3 m	4 m

9 The cost (£C) of producing N copies of a poster is given by $C = 4 + \frac{N}{100}$. How much would it cost to produce the following numbers of copies?

    a 300 b 400 c 100 d 500 e 1000 f 50

Example

If $N = 200$, $C = 4 + \frac{200}{100} = 4 + 2 = £6$

10 The area (A) of this shape is given by this formula $A = \frac{1}{2}H(T + B)$.

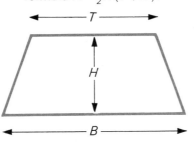

Find the areas of these shapes.

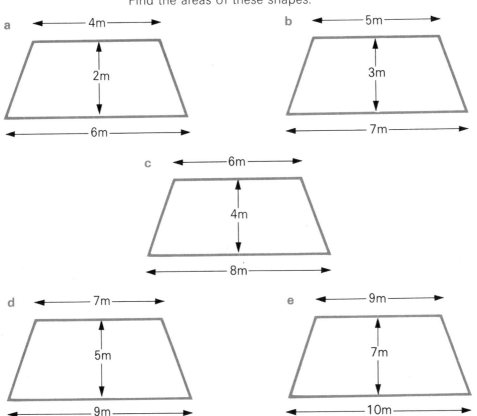

unit 12 Percentages

If 100 pupils enter an examination and 57 pass, we say that 57 per cent pass. It is usually written like this: 57%.

If the Government says that the unemployment rate is 6% they mean that averaged over the whole country, 6 out of every 100 men are unemployed.

per cent or % means 'out of every hundred'

If in one school 56% pass the English examination and 63% pass the mathematics examination we can see that the pass rate is better in mathematics than in English.

Suppose that 200 enter the mathematics examination and 120 pass, and 50 enter the English examination and 40 pass. Which has the better pass rate, mathematics or English?
One way of doing a problem like this is to change both the pass rates into percentages.

Mathematics: What is 120 out of 200 as a percentage?
If we halve each number we get 60 out of 100.
        Answer 60%

English: What is 40 out of 50 as a percentage?
If we double each number we get 80 out of 100.
        Answer 80%   English has the higher pass rate.

Example

Write these as percentages.
a   7 out of 100   b 12 out of 200
c 24 out of 300   d 30 out of 50
e   5 out of 25   f   7 out of 20
g   8 out of 10   h 40 out of 400

a 7%
b Divide each number by 2 to get 6 out of 100   6%
c Divide each number by 3 to get 8 out of 100   8%
d Multiply each number by 2 to get 60 out of 100 60%
e Multiply each number by 4 to get 20 out of 100 20%
f Multiply each number by 5 to get 35 out of 100 35%
g Multiply each number by 10 to get 80 out of 100 80%
h Divide each number by 4 to get 10 out of 100 10%

Exercise 12.1

Write each of these as a percentage.

	a	b
1	6 out of 100	8 out of 200
2	3 out of 50	7 out of 25
3	16 out of 200	56 out of 100
4	10 out of 20	5 out of 20
5	60 out of 300	68 out of 200
6	17 out of 200	$2\frac{1}{2}$ out of 50

unit 12 / page 85

	c	d
1	9 out of 300	16 out of 400
2	5 out of 20	3 out of 10
3	15 out of 50	2 out of 25
4	1 out of 10	50 out of 500
5	17 out of 50	20 out of 25
6	$3\frac{1}{2}$ out of 25	3 out of 4

7 In a school 200 pupils enter for examinations in Mathematics, English, Science, and French. 140 pass in Mathematics, 120 pass in English, 80 pass in Science, and 110 pass in French.
   a What percentage pass in Mathematics?
   b What percentage pass in English?
   c What percentage pass in Science?
   d What percentage pass in French?

8 Boys from two classes, 2A and 2B went for a swimming test. In 2A 15 passed out of 20. In 2B 18 passed out of 25.
   a What percentage passed in 2A?
   b What percentage passed in 2B?
   c Which class did best?

9 Two types of tyre are tested by driving them for 20 000 miles. 50 Roadmaster tyres are tested and 13 wear out. 25 Safedrive tyres are tested and 5 wear out.
   a What percentage of Roadmaster tyres wear out?
   b What percentage of Safedrive tyres wear out?
   c Which type of tyre did best in the test?

10 200 pupils took an examination. 120 passed, and 80 failed.
   a What percentage passed?
   b What percentage failed?
   c Add together the answers to a and b.

11 50 pupils took an examination. 40 passed, and 10 failed.
   a What percentage passed?
   b What percentage failed?
   c Add together the answers from a and b.

12 Of pupils who took an examination 70% passed. What percentage failed?

13 Of pupils who took an examination 80% passed. What percentage failed?

14 Of the pupils in a school 72% are boys. What percentage are girls?

page 86/unit 12

15 Of the families in a town 17% have two or more cars. 53% of the families have only one car. What percentage of the families have no car?

16 Two types of carpet are tested on a machine.
Twenty-five pieces of Longlife are tested and three wear out.
Twenty pieces of Extralux are tested and two wear out.

    a What percentage of Longlife wear out?
    b What percentage of Extralux wear out?
    c Which seems to be the better carpet?

Finding percentages of numbers

If 6% of the men in this country are unemployed, how many might you expect to be unemployed out of 100?

Since 6% means 6 out of every hundred the answer is obviously **6 men**.
    What we have done is to find 6% of 100.

How many would you expect to be unemployed out of 200?

Six out of every 100 are unemployed, so if we double each number we get: 12 out of 200. The answer is **12 men**.
    What we have done is to find 6% of 200.

How many would you expect to be unemployed out of 50?
6 out of every 100 are unemployed, so if we halve each number we get: 3 out of 50. The answer is **3 men**.
    What we have done is to find 6% of 50.

Example
Find 13% of 200.
13 out of 100 is the same as 26 out of 200 (Double each number). **Answer 26.**

Example
Find 8% of 50.
8 out of 100 is the same as 4 out of 50 (Halve each number). **Answer 4.**

Example
Find 40% of 25.
40 out of 100 is the same as 10 out of 25 (Divide each number by four). **Answer 10.**

Example
A factory which produces tyres expects 4% of them to be rejected because of faults. How many would you expect to be rejected out of 700?
What we have to do is to find 4% of 700.
4 out of 100 is the same as 28 out of 700 (Multiply each number by 7). **Answer 28 tyres.**

Exercise 12.2

	a	b	c	d	e	f
1 Find 20% of	100	200	50	25	20	10
2 Find 40% of	100	200	50	25	20	10
3 Find 8% of	50	200	100	300	400	25
4 Find 30% of	50	20	10	200	300	500
5 Find 45% of	20	100	200	300	400	500
6 Find 80% of	50	25	20	10	5	40

7 A factory which produces tyres expects 4% of them to be rejected because of faults.
   a How many would be rejected out of 100 tyres?
   b How many would be rejected out of 50 tyres?
   c How many would be rejected out of 25 tyres?
   d How many would be rejected out of 200 tyres?

8 A school finds that 20% of the pupils they put in for an examination usually fail.
   a How many would you expect to fail out of 100?
   b How many would you expect to fail out of 50?
   c How many would you expect to fail out of 25?
   d How many would you expect to fail out of 200?
   e How many would you expect to fail out of 300?

9 A hotel adds a 10% service charge to every bill.
   a What would be the service charge on a £100 bill?
   b What would be the service charge on a £50 bill?
   c What would be the service charge on a £20 bill?
   d What would be the service charge on a £200 bill?

10 An alloy is made of 40% copper, and 60% tin.
   a What weight of each would there be in 100 kg?
   b What weight of each would there be in 50 kg?
   c What weight of each would there be in 25 kg?
   d What weight of each would there be in 20 kg?
   e What weight of each would there be in 10 kg?

11 A carpet firm gives a discount of 4% on bills which are paid within 14 days.
   a What would be the discount on a £100 bill?
   b What would be the discount on a £50 bill?
   c What would be the discount on a £25 bill?
   d How much would you have to pay on a £100 bill with the discount?
   e How much would you have to pay on a £50 bill with the discount?
   f How much would you have to pay on a £25 bill with the discount?

Writing one number as a percentage of another

If 2 pupils out of 7 fail a driving test, what percentage is this? You will see that we have to write 2 as a percentage of 7.

If we try to use the method shown at the start of this unit you will see that it will not work, and we have to use the following method.

1 Write the number that pass as a fraction. → $\frac{2}{7}$

2 Change this fraction into a decimal by dividing 2 by 7 to get an answer of 0·28. (see p. 52)

```
  0·28
7│200
   14↓
   60
   56
```

3 Remember that $0\cdot 28 = \frac{28}{100} = 28\%$.

i.e. 28% fail the test.

Example

A boy gets 5 out of 9 marks on a test. What percentage is this?

$$5 \text{ out of } 9 = \frac{5}{9} = 0\cdot 555 = \frac{55\cdot 5}{100} = 55\cdot 5\%$$

```
   0·555
9│ 5000
   45↓
   50
   45↓
   50
   45
    5
```

You will see that in this case we found the answer to three figures, which is accurate enough for most purposes.

Exercise 12.3

	a	b	c	d	e
1 Write 4 as a percentage of	5	6	7	8	9
2 Write 5 as a percentage of	6	7	8	9	11
3 Write 6 as a percentage of	7	8	9	11	13
4 Write 9 as a percentage of	14	12	22	43	57
5 Write 37 as a percentage of	56	68	94	75	158

6 In 1972 9 pupils took an examination and 7 passed.
In 1973 8 pupils took an examination and 6 passed.

a What percentage passed in 1972?
b What percentage passed in 1973?
c Which year had the best results?

7 In 1972 a teacher put 8 pupils in for an

examination and 7 passed. The next year he decided that if he doubled the homework he would get better results. In 1973 he put 9 pupils in for the examination and 7 passed.

a What percentage passed in 1972?

b What percentage passed in 1973?

c Did doubling the homework improve the examination results?

8 Two classes had a competition to see which of them did best in athletics. In class 3H, 2 out of 9 failed the test. In class 3G, 1 out of 5 failed the test.

a What percentage failed in 3H?

b What percentage failed in 3G?

c Which class did best?

9 Two types of shoe are tested on a machine. 24 pairs of Extratuff are tested and 7 wear out. 35 pairs of Everlast are tested and 10 wear out.

a What percentage of Extratuff wear out?

b What percentage of Everlast wear out?

c Which seems to be the strongest shoe?

10 During a period of price restraint, a government says that prices must not increase by more than 5%. Which of the following prices have been increased by more than 5%?

a From 52p to 55p.

b From 75p to 78p.

c From £1·23 to £1·25.

11 Which of the following is the greater percentage error.

a An error of 1 m in measuring 26 m?

b An error of 3 cm in measuring 62 cm?

12 A cheap weighing machine is claimed to weigh with an accuracy of 1%. Weights of 5 g, 15 g, and 30 g are weighed and the machine gives the weights as 5·06 g, 15·16 g, 30·2 g. Is the claim accurate?

13 A washing machine is priced at £78. It is also possible to buy it in 12 instalments of £7·50 paid over one year.

a What is the total cost if the machine is bought in instalments?

b How much more is paid if the machine is bought this way?

c What rate of interest is being charged? (The rate of interest is obtained by working out the answer to b as a percentage of the original price.)

14 A car is priced at £983. It is also possible to buy it in 12 monthly instalments of £90.
  a What is the total cost if the car is bought in instalments?
  b How much more is paid if the car is bought this way?
  c What rate of interest is being charged? (The rate of interest is obtained by working out the answer to **b** as a percentage of the original price.)

Finding percentages of numbers (harder examples)

If a hotel adds a 5% service charge onto every bill, what will be the service charge on a bill of £72?

The method we were using in the first part of this Unit will not work very well in this case, and another method is needed for all problems of this type.

5% of £72 means $\frac{5}{100} \times 72$ which can be written 0.05×72 which comes to £3·60.

Example
Find 8% of £27.   Find 27% of 8 kg.
$0.08 \times 27 = £2.16$ $0.27 \times 8 = 2.16$ kg

To deal with the most common fractional percentages remember that: $\frac{1}{2} = 0.5$, $\frac{1}{4} = 0.25$, $\frac{3}{4} = 0.75$, $\frac{1}{3} = 0.33$, $\frac{2}{3} = 0.67$

Example
$4\frac{1}{2}\% = 4.5\%$ so we would multiply by 0·045.
$17\frac{2}{3}\% = 17.67\%$ so we would multiply by 0·1767.

If we had to find say $5\frac{2}{7}\%$ we would divide 2 by 7 to get 0·28 so $5\frac{2}{7}\% = 5.28\%$ and we would multiply by 0·0528.

Example
A man invests £350 in a Building Society at $6\frac{3}{4}\%$. How much interest would he get every year?
$6\frac{3}{4} = 6.75$, so Interest $= 0.0675 \times 350 =$
$£23.625 = £23.62\frac{1}{2}$.

Exercise 12.4

		a	b	c	d	e
1	Find 5% of	£7	£9	£6	£3	£8
2	Find 23% of	£3	£5	£7	£8	£9
3	Find 7% of	32 kg	47 kg	59 kg	29 kg	84 kg
4	Find $5\frac{1}{2}\%$ of	£7	£9	£4	£2	£5
5	Find $7\frac{3}{4}\%$ of	8 m	9 m	16 m	35 m	8·4 m
6	Find $2\frac{1}{3}\%$ of	£3·24	£4·85	£23·67	£0·78	34p

unit 12/page 91

7 A firm gives a discount of 6% on bills that are paid within 21 days.
  a What is the discount on a bill of £34?
  b What is the discount on a bill of £24?
  c What is the discount on a bill of £3·56?

8 A firm gives a discount of 4% on bills that are paid within 28 days. What is the bill in each of the following cases after the discount has been taken off?
  a £25 **b** £57 **c** £247 **d** £23·56 **e** £0·89

9 A firm producing light bulbs finds that about 3% of their output is defective.
  a How many defective bulbs would they expect to find in 300?
  b How many defective bulbs would they expect to find in 500?
  c How many non-defective bulbs would they expect to get in 700?
  d How many non-defective bulbs would they expect to get in 2200?

10 An alloy is made of 23% tin, 45% copper, and 32% zinc.
  a What weight of each would be required to make 70 kg of alloy?
  b What weight of each would be required to make 15 kg of alloy?
  c What weight of each would be required to make 240 kg of alloy?
  d What weight of each would be required to make 240 tonnes of alloy?

11 A salesman is paid commission of 8% on his sales. What commission will he be paid on sales of:
  a £70 **b** £260 **c** £463 **d** £3500 **e** £3743?

12 A salesman is paid commission of $7\frac{3}{4}$% on his sales. What commission will he be paid on sales of:
  a £56 **b** £240 **c** £346 **d** £2500 **e** £2357?

unit 13 Indices

What is the area of this square of side s?

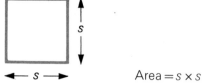

Area $= s \times s$

What is the volume of this cube?

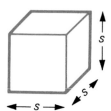

Volume $= s \times s \times s$

Formulae like this come up often in mathematics and there is a shorter way of writing them; like this:

Area $= s^2$

Volume $= s^3$

The small number is called an **index** and it tells you how many of the same letters or numbers have to be multiplied together. Here are some more examples of indices. (If you have more than one index they are called indices.)

$a \times a \times a \times a = a^4$

$p \times p \times p = p^3$

$c^2 = c \times c$

$y^4 = y \times y \times y \times y$

You should remember that a^2 is not the same as $2a$.

$a^2 = a \times a$
$2a = a + a$

b^3 is not the same as $3b$.

$b^3 = b \times b \times b$
$3b = b + b + b$

When you read s^2 you can either say 's to the power of two', or 's squared'.

When you read s^3 you can either say 's to the power of three', or 's cubed'.

s^4 and s^5 are called 's to the power of four', and 's to the power of five', and so on.

unit 13/page 93

Exercise 13.1

Write the examples underneath using indices.

a
$b \times b \times b = b^3$
1. $d \times d \times d \times d$
2. $h \times h \times h$
3. $k \times k$
4. $g \times g \times g \times g \times g$
5. $w \times w \times w$
6. $c \times c \times c \times c$
7. $r \times r$
8. $y \times y \times y \times y \times y$
9. r

Write the examples underneath without using indices.

b
$n^2 = n \times n$
k^3
h^2
d^5
e^3
u^4
s^2
w^5
p^1
w^6

Formulae with indices

Example

Find the area of this square.

Area $= 5^2 = 5 \times 5 =$ **25 sq cm**

Example

Find the volume of this cube.

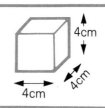

Volume $= 4^3 = 4 \times 4 \times 4$
$=$ **64 cubic cm**

Example

Find the value of 7^2.
$7^2 = 7 \times 7 =$ **49**

Example

Find the value of 2^3.
$2^3 = 2 \times 2 \times 2 =$ **8**

Example

Find the value of $3^2 + 4^3$.
$3^2 + 4^3 = 3 \times 3 + 4 \times 4 \times 4 = 9 + 64 =$ **73**

If $a = 4$, $b = 5$, $c = 1$, find the value of (i) a^2, (ii) $a^2 + b^2$, (iii) $b^2 - c^2$, (iv) $a^3 - b^2$, (v) $4b^2$.

    (i) $a^2 = 4 \times 4 =$ **16**
   (ii) $a^2 + b^2 = 4 \times 4 + 5 \times 5 = 16 + 25 =$ **41**
  (iii) $b^2 - c^2 = 5 \times 5 - 1 \times 1 = 25 - 1 =$ **24**
  (iv) $a^3 - b^2 = 4 \times 4 \times 4 - 5 \times 5 = 64 - 25 =$ **39**
   (v) $4b^2 = 4 \times 5 \times 5 = 4 \times 25 =$ **100**

page 94/unit 13

Example → Find the value of 1^4 and 0^3.
$1^4 = 1 \times 1 \times 1 \times 1 = \mathbf{1}$
$0^3 = 0 \times 0 \times 0 = \mathbf{0}$

Exercise 13.2

1 Find the areas of these squares.

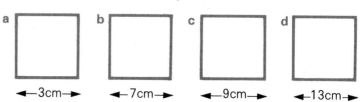

a ◄—3cm—►   b ◄—7cm—►   c ◄—9cm—►   d ◄—13cm—►

2 Find the volumes of these cubes.

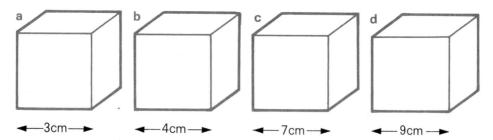

a ◄—3cm—►   b ◄—4cm—►   c ◄—7cm—►   d ◄—9cm—►

Find the value of the following:

	a	b	c	d
3	2^1	2^2	2^3	2^4
4	3^1	3^2	3^3	3^4
5	5^1	5^2	5^3	5^4
6	2^2+1^2	2^2+3^2	4^2+3^2	4^2+5^2
7	4^2+7^2	6^2+5^2	8^2+7^2	13^2+17^2
8	5^2-4^2	4^2-3^2	3^2-2^2	2^2-1^2
9	7^2-4^2	6^2-5^2	8^2-7^2	16^2-11^2
10	4^2+2^4	5^3+2^4	1^5+5^1	$2^2+3^3+4^4$

If $p=1$, $q=2$, $r=5$, $s=7$, find the value of the following:

	a	b	c	d
11	r^2	s^2	q^2	p^2
12	q^3	p^3	r^3	s^3
13	p^2+q^2	q^2+r^2	r^2+s^2	q^2+s^2
14	r^2-q^2	r^2-p^2	s^2-q^2	s^2-p^2
15	$3q^2$	$2p^2$	$3r^2$	$4s^2$
16	$2q^2+3p^2$	$4r^2-2q^2$	r^q	$2p^2q^2$

unit 14 Rectangular Numbers, Prime Numbers, and Factors

If you have 6 dots they can be arranged in a rectangle like this.

If you have 18 dots they can be arranged in a rectangle in two ways.

A number which can be expressed as a rectangle of dots is called a **rectangular number**. It is also called a **composite number**.

Can you arrange 7 dots in a rectangle?
Can you arrange 11 dots in a rectangle?

You will see that there are some numbers which are not rectangular numbers. They are called **prime numbers**.
7 and 11 are prime numbers.

1 is not regarded as a prime number as it can be fitted into a rectangle like this.

Since 6 dots can be arranged in one rectangle you will see that $6 = 2 \times 3$.
Since 18 dots can be arranged in a rectangle in two ways we may write either $18 = 2 \times 9$ or $18 = 3 \times 6$.

Example

Which of these numbers is a prime number? Which of these numbers is a rectangular number? If the number is a rectangular number show all the ways it may be arranged as a rectangle.

a 16   b 13

a 16 is a rectangular number. **$16 = 2 \times 8$** or **$16 = 4 \times 4$**
b 13 is a prime number.

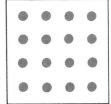

Exercise 14.1

Which of these numbers is a prime number? Which of these numbers is a rectangular number. If the number is a rectangular number show all the ways it can be arranged as a rectangle. Example: $18 = 2 \times 9 = 3 \times 6$.

	a	b	c	d	e
1	3	9	5	8	10
2	4	14	12	17	15
3	19	21	20	30	24
4	22	25	23	27	33
5	26	31	28	32	29
6	43	36	39	34	37
7	35	40	38	41	44
8	42	64	72	97	101
9	60	107	81	91	93
10	100	67	144	121	113

Finding the factors of numbers

We may write 18 as $18 = 3 \times 6$, but since $6 = 2 \times 3$, we may write $18 = 2 \times 3 \times 3$.

You will see that we have now written 18 as a set of prime numbers multiplied together. We have written 18 as a product of prime numbers.

The numbers that divide into 18 are 1, 2, 3, 6, 9, 18. These are called the factors of 18. We may write

Factors of $18 = \{1, 2, 3, 6, 9, 18\}$

Example

Write each of these numbers as a product of prime numbers, and also write down the set of their factors.
a 12   b 15

a $12 = 2 \times 2 \times 3$ $\{1, 2, 3, 4, 6, 12\}$
b $15 = 3 \times 5$ $\{1, 3, 5, 15\}$

Exercise 14.2

Write each of these numbers as a product of prime numbers, and write down the set of their factors.

	a	b	c	d	e
1	4	6	9	16	20
2	21	22	24	25	28
3	30	32	33	34	35
4	36	40	45	48	72
5	64	70	76	100	144

The answers to some of the above examples can be written in a shorter way using indices. (See Unit 13.)

Example

$$18 = 2 \times 3 \times 3 = 3^2 \times 2$$
$$150 = 2 \times 3 \times 5 \times 5 = 2 \times 3 \times 5^2$$

Do the examples in Exercise 14.2 which may be written in a shorter way using indices.

Finding prime numbers

One way of finding prime numbers was discovered by a mathematician called Eratosthenes who lived over 2000 years ago. We will use his method to find all the primes from 1 to 50.

First we write down all the numbers from 1 to 50.
Then we cross out all the numbers that 2 will divide into. Every second number. We cross out 4, 6, 8 and so on.

Then we cross out all the numbers that 3 will divide into. Every third number. We cross out 6, 9, 12 and so on.

We do not need to cross out the numbers that 4 will divide into because they will have been crossed out with the numbers that 2 divides into.

We then need to cross out the numbers that 5 and 7 divide into.
We do not need to cross out the numbers that 6 divides into as these will have been crossed out when we crossed out the numbers that 2 and 3 divide into.
We do not need to go further than 7. Can you see why? This will be explained later.

You will see that the prime numbers up to 50 are:
2, 3, 5, 7, 11, 13, 17, 19, 23, 29, 31, 37, 41, 43, 47.

~~1~~ 2 3 ~~4~~ 5 ~~6~~ 7 ~~8~~ ~~9~~ ~~10~~
11 ~~12~~ 13 ~~14~~ ~~15~~ ~~16~~ 17 ~~18~~ 19 ~~20~~
~~21~~ ~~22~~ 23 ~~24~~ ~~25~~ ~~26~~ ~~27~~ ~~28~~ 29 ~~30~~
31 ~~32~~ ~~33~~ ~~34~~ ~~35~~ ~~36~~ 37 ~~38~~ ~~39~~ ~~40~~
41 ~~42~~ 43 ~~44~~ ~~45~~ ~~46~~ 47 ~~48~~ ~~49~~ ~~50~~

Exercise 14.3

1 Find all the prime numbers between 50 and 100. You will need to cross out the numbers that can be divided by 2, 3, 5, 7.

2 Find all the prime numbers between 100 and 150. You will need to cross out the numbers that can be divided by 2, 3, 5, 7, 11.

3 Find all the prime numbers between 500 and 550. You will need to cross out the numbers that can be divided by 2, 3, 5, 7, 11, 13, 17, 19, 23.

Finding out if a number is prime

There is no easy way of finding out if a number is prime. The usual way is to see if there is a smaller prime that divides into it.

Example

Is 91 a prime number?

Divide by 2. $91 \div 2 = 45$ remainder 1
Divide by 3. $91 \div 3 = 30$ remainder 1
Divide by 5. $91 \div 5 = 18$ remainder 1
Divide by 7. $91 \div 7 = 13$

91 is not a prime number as it can be written $91 = 7 \times 13$.

You should notice that we do not need to divide by 4 because if 4 divides into 91, then 2 will, and we have tried 2 already.
Neither do we need to try 6 because if 6 goes, then 2 and 3 will, and we have tried these already.

Example

Is 97 a prime number?

Divide by 2. $97 \div 2 = 48$ r 1
Divide by 3. $97 \div 3 = 32$ r 1
Divide by 5. $97 \div 5 = 19$ r 2
Divide by 7. $97 \div 7 = 13$ r 6
Divide by 11. $97 \div 11 = 8$ r 9

There is no need to go any further since the answer we get on dividing is less than the number we are dividing by. If for example we thought that the next prime number 13 might go exactly we would get

$97 \div 13 =$ a number less than 13.

This is impossible as we have already tried all the numbers less than 13, and they do not go exactly.

97 is a prime number.

To find out if a number is a prime number, divide by all the primes starting with 2 until either you get a number that goes exactly, or until the answer you get is less than the number you are dividing by.
In practice you do not need to divide by 2 as no even number can be a prime, and you do not need to divide by 5 as any number that can be divided by 5 ends in 5 or 0.

Exercise 14.4

Find out which of these numbers are primes. If a number is not a prime, write down the smallest number that goes into it.

	a	b	c	d	e
1	101	129	109	208	113
2	139	151	355	173	213
3	181	217	233	237	317
4	247	419	643	221	169
5	1373	391	529	1447	899

unit 15 Number Pictures and Averages

Three boys stand at the school gate for half an hour and count the number of cars, buses and lorries that pass. They record the numbers like this:

Cars   ⊬⊬⊬ //
Buses   ⊬⊬⊬
Lorries   ⊬⊬⊬ /

Every fifth stroke crosses through the previous four. You will see that the number of cars was 7.

How many buses did they count?
How many lorries did they count?

Pictograms

One way of showing the number of buses would be to draw a set of pictures. This is called a pictogram, and a pictogram of the boys' results is shown below.

Two girls are sent to count the number of men and women that go into the public library during half an hour. Here are their results.

Men   ⊬⊬⊬ ⊬⊬⊬ ⊬⊬⊬ ⊬⊬⊬ ⊬⊬⊬ ⊬⊬⊬ ⊬⊬⊬ ⊬⊬⊬ ⊬⊬⊬ ⊬⊬⊬ ⊬⊬⊬ ⊬⊬⊬ ⊬⊬⊬ //

Women   ⊬⊬⊬ ⊬⊬⊬ ⊬⊬⊬ ⊬⊬⊬ ⊬⊬⊬ ⊬⊬⊬ ⊬⊬⊬ ⊬⊬⊬ ⊬⊬⊬ ////

How many men went into the library?
How many women went into the library?

A pictogram showing these results would be rather large. We can make it smaller if we use one picture to stand for 5 men or women.

You will see that part of a picture stands for less than five.

Exercise 15.1

1 This is a record of the eye colours of children in a class:

Brown //// //// //
Blue ////
Hazel //// //
Grey ///
Black ///
Green //

    a What is the most common colour?
    b How many children have hazel eyes?
    c How many children have brown eyes?
    d How many children are there in the class?

2 The pictogram shows how many boys prefer soccer and how many prefer rugby.

    a Which game is the more popular?
    b How many boys prefer rugby?
    c How many boys prefer soccer?

3 Five girls stand at the school gate recording the colours of passing cars. Here is a pictogram of their results:

    a How many white cars were there?
    b How many cars were not black?
    c How many cars passed the gate altogether?

4 A survey was made of how the children in a school travel there each day. Here is a pictogram showing the results:

= 10 children

a How many children walk to school?
b How many travel by bus?
c How many cycle to school?
d How many children go to school by car?
e How many travel by train?

5 Here is a pictogram showing how much pocket money each of three children get:

● = 10p

a How much does Bill get?
b How much does Mary get?
c How much does Susan get?

6 John counted the different kinds of birds on the bird table in his garden. Using a simple diagram to represent one bird, draw a pictogram of the numbers shown in the table.

kind of bird	number present
blackbirds	3
sparrows	4
chaffinches	1
starlings	6

7 The times given to different types of programme on a television channel during a particular week is given in the table below.

type of programme	hours of television
comedy	20
westerns	15
sport	10
news	5

Using 📺 to represent five hours of television time draw a pictogram to illustrate the table.

Bar charts

Instead of using pictograms we may use **bar charts**. The bar charts we could draw instead of the two pictograms on page 99 are shown below.

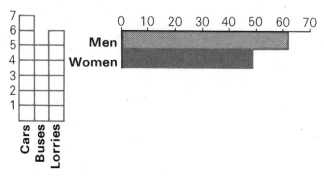

You should notice that we can draw the bars horizontally or vertically. In the first bar chart a square has been used for each car, bus, and lorry, in the second the number of men or women has to be read off along the top of the chart.

Exercise 15.2

1 The bar chart shows the number of different flavoured milk shakes sold during one lunch hour. Which was the most popular flavour?

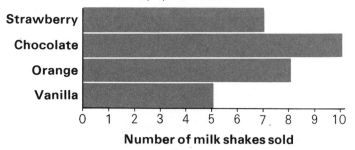

a How many strawberry milk shakes were sold?

b How many chocolate and vanilla milk shakes were sold?

c How many milk shakes were sold altogether?

2 Draw a bar chart similar to that in question 1 and using the following information.

Number of milk shakes sold in one lunch hour

Flavour	Vanilla	Chocolate	Strawberry
Number	4	7	8

Flavour	Lime	Orange	Lemon
Number	3	6	5

3 The number of pupils away in a class is shown on the bar chart below.

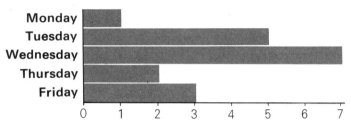

a How many pupils were away on Monday?
b How many pupils were away on Friday?
c What day had the most number of pupils away?
d What day had the least number of pupils away?
e What was the total of absences for the week?
f Is the total number of pupils away during the week the same as the answer to e? Give reasons for your answer.

4 Draw a bar chart similar to that in question 3 and using the following information.

Day	Monday	Tuesday
Number of pupils away	7	6

Day	Wednesday	Thursday
Number of pupils away	3	0

Day	Friday
Number of pupils away	5

5 The number of road accidents in a town during one week is shown on the bar chart below.

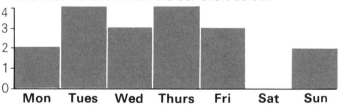

a What day had the highest number of accidents?
b What day had the lowest number of accidents?
c What days had the same number of accidents?
d How many accidents were there on Monday?
e How many accidents were there during the week?

6 Draw a bar chart similar to that in question 5 and using the following information.

Day	Monday	Tuesday
Number of accidents	4	6

Day	Wednesday	Thursday
Number of accidents	5	7

Day	Friday	Saturday	Sunday
Number of accidents	2	2	3

7 Here is a list of the marks obtained by twenty-five boys in a test: 6, 5, 5, 5, 7, 9, 2, 6, 1, 7, 4, 10, 9, 9, 1, 10, 2, 5, 8, 3, 8, 5, 2, 10, 9. Copy and complete this table:

Mark	1	2	3	4	5	6	7	8	9	10
Number of boys						2				

Draw a bar chart illustrating the distribution of the marks.

8 The bar chart shows the records of rainfall for a school weather station at monthly intervals throughout 1972. Use the bar chart to complete the following table.

Month	J	F	M	A	M	J	J	A	S	O	N	D
Millimetres of rainfall	100					85						

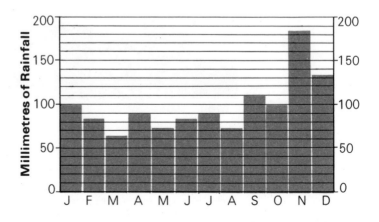

Averages

A class has ten boys and ten girls. They are given a test and the marks are shown below. Who did better, the boys or the girls?

Boys	7	5	3	6	8	4	6	2	4	8
Girls	2	9	3	8	4	7	5	6	4	8

One way of finding out is to add all the marks together. The girls' marks add up to 56; the boys' marks add up to 53. You will see that the girls did better.

The problem of finding out who did best is more difficult if there are a different number of boys and girls.

| Boys | 7 | 8 | 5 | 4 | 2 | 7 | 6 | 9 | — | — | Total of boys' marks = 48 |
| Girls | 4 | 2 | 5 | 2 | 8 | 6 | 7 | 2 | 9 | 5 | Total of girls' marks = 50 |

The best way of comparing the marks of the boys and girls is to see how many marks the girls would have if their marks were spread equally amongst the girls, and do the same with the boys' marks. This is called the average mark.

Average mark for the boys = 48 ÷ 8 = 6.
Average mark for the girls = 50 ÷ 10 = 5.

You will see that in this case the boys did slightly better.

Example

Find the average of 5, 6, 4, 8, 9, 3, 7, 2, 8.

The sum of the numbers is 52.

There are nine numbers

Average = 52 ÷ 9 = 5·77.

$$9 \overline{)52 \cdot 00} \\ 45 \\ 70 \\ 63 \\ 70$$

You will see that the average does not come out exactly and it has been worked out to three figures. This is accurate enough for most purposes.

Exercise 15.3

1 Calculate the averages of the following sets of numbers:

a 6, 3, 8, 7
b 9, 4, 7, 12
c 7, 1, 6, 2, 2, 0
d 2, 5, 6, 4, 4, 6, 8
e 11, 83, 78, 74, 34
f 90, 97, 26, 77, 51, 57, 92
g 76, 27, 10, 75
h 79, 97, 49, 24, 31

2 Three boys spent a day fishing; John caught two fish, Michael caught five and Peter eleven. What was the average number of fish caught by the three boys?

3 Mary spent 3p on sweets and Ann spent 6p; what was the average amount spent by the two girls?

4 The temperatures at twelve noon during one week in April are shown below. What was the average noonday temperature?

Day	S	M	T	W	T	F	S
Temperature Degrees C	15	14	16	16	17	15	12

5 Three children have heights of 1·46 m, 1·52 m, and 1·58 m. What is their average height?

6 The price of the same model of a washing machine in five different shops was £98, £103, £107, £105, and £112. What was the average price of the washing machine?

7 Ten pupils took a biology exam and their marks were as follows: 48%, 51%, 27%, 64%, 72%, 95%, 63%, 55%, 73%, and 62%. What was their average mark?

8 Team A in a tug-of-war consists of men of weights of 75 kg, 80 kg, 83 kg, and 98 kg. The four members of team B have an average weight of 82 kg. What is the average weight of team A? Which team is the heavier?

9 Find the average of the following sets of numbers:

　a 7, 9, 9, 10, 11, 11, 13, 16, 17;

　b 29, 75, 43, 11, 0, 12, 34;

　c 33, 59, 64, 57, 28, 78, 16, 32;

　d 4, 4, 5, 8, 5, 8, 4, 1, 0;

　e 11, 13, 13, 9, 8, 17, 16, 4, 15, 13, 18, 14, 15.

10 A man's monthly earnings during the last year were £143, £147, £153, £109, £115, £138, £142, £147, £146, £149, £153 and £150. What were his average monthly earnings?

11 In a test the boys' marks were 7, 5, 7, 8, 4, 3, 8, 9, 10, 9, 7, 6, and 5. The girls' marks were 8, 6, 5, 7, 9, 4, 7, 10, 6, 8, and 7.

　a What was the average mark for the girls?

　b What was the average mark for the boys?

　c Did the girls or boys do better in the test?

　d What was the average mark for the boys and girls combined?

12 Mary weighed a dozen apples and found the weights to be 93, 87, 96, 82, 98, 95, 102, 75, 86, 94, 88, and 96 grammes. What was the average weight of the apples?

Pie charts

A man earns £16 every week. He spends £1 on fares, £2 on clothes, £4 on food, £3 on entertainments, £4 on rent. He saves £2. This could be shown on a bar chart. It could also be shown on a pie chart. Both ways of doing this are shown.

How a man spends his weekly wage of £16

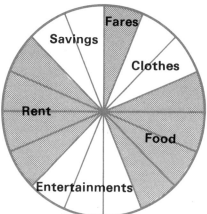

Exercise 15.4

1 The pie chart shows how a man spends his weekly wage of £16. How much does he spend on:

a Fares?
b Entertainments?
c Food?
d Clothes?
e Rent?
f Savings?

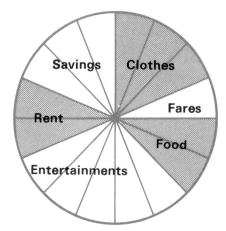

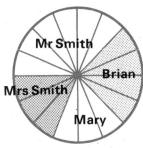

2 The pie chart shows how a family spend their weekly income of £32. How much do they spend on:
  a Rent?
  b Clothes?
  c Savings?
  d Entertainment?
  e Food?
  f Fares?

3 The pie chart shows how much each member of the Smith family earns. If Brian earns £16, how much does:
  a Mr Smith earn?
  b Mrs Smith earn?
  c Mary Smith earn?
  d The whole family earn?

For each of the next three questions you will need to draw a circle divided into ten equal parts like the one below. (Each of the parts will represent 10%.) It can be copied using tracing paper.

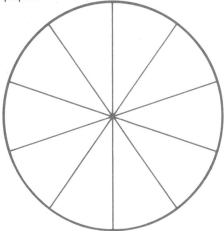

4 The following figures show how a man spends his income. Savings 10%, rent 40%, food 20%, fares 20%, other items 10%. Draw a pie chart to illustrate this.

5 The following figures show how a Council spends the rates. Education 40%, housing 20%, roads 10%, other items 30%. Draw a pie chart to illustrate this.

6 The following figures show how a family spends its income. Entertainments 10%, savings 10%, clothes 20%, rent 20%, food 30%, fares 10%. Draw a pie chart to illustrate this.

7 The pie chart shows how a man spent his money during a holiday. He spent £6 on his fares. How much did he spend:

a on his hotel bill? c altogether?

b on hiring the boat?

8 John, Mary and Joan have £1·20 between them; from the pie chart say how much each has.

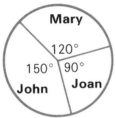

9 In a pet shop there are three dogs, five cats, six tortoises and two rabbits. Draw a circle and divide it into sixteen equal sectors as shown in the figure. Use this to make a pie chart showing how many of each kind of animal there are in the shop.

10 Richard has twice as much money as Susan and Mary each have, and half as much as Irene. Draw a pie chart showing how the money is divided between them.

11 Collect information from pupils in your class on any of the following and make pictograms, bar charts or pie charts to illustrate them:

Favourite football team

Favourite author

Favourite pop group

Favourite film or T.V. star

Ways of travelling to school

Time spent watching television

Primary school attended

Favourite hobby or sparetime activity

Time spent on each school subject

Favourite magazine

Favourite T.V. programme

Any other topic of your choice.

unit 16 Equations

Here is an example of an equation:
$$a+7=10$$
For this to make sense you can see that a must be equal to 3, since $3+7=10$.

Finding the value of a letter in an equation is called solving the equation.

Here is another example.
$$b-2=6$$
For this to make sense you can see that b must be equal to 8, since $8-2=6$.

Many problems in mathematics are solved using equations. In this Unit you will first learn how to solve equations, and then you will learn how to solve problems using equations.

Example

Solve the equation $x+7=12$.
You can see that $x=5$ and this is got by taking 7 from twelve. So we can write the solution like this.
$$x+7=12$$
$$\Rightarrow x = 12-7$$
$$\Rightarrow x = 5$$

Example

Solve the equation $x+26=49$ we can say:
$$\Rightarrow x = 49-26$$
$$\Rightarrow x = 23$$

Example

Solve the equation $v-4=8$.
You can see that v must be 12 and this is got by adding 4 to 8. So we can write the solution like this.
$$v-4 = 8$$
$$\Rightarrow v = 8+4$$
$$\Rightarrow v = 12$$

Example

Solve the equation $v-34=47$
$$\Rightarrow v = 47+34$$
$$\Rightarrow v = 81$$

If you look at the last four examples you can see that if a number is moved from one side of the = sign to the other we must change the sign in front of it. That is, + changes into −, and − changes into +. This is often remembered as follows.

change the side
change the sign

unit 16/page III

Example

Solve the equation $a - 36 = 69$ and check the answer.

$$a - 36 = 69$$
$$\Rightarrow a = 69 + 36$$
$$\Rightarrow a = 95$$

To check this answer we put it back into the equation
$$a - 36 = 95 - 36 = 59$$

Something is wrong here as this should come to 69. If we look back at the last line of the equation we can see that it should be
$$a = 105 \quad \text{and } not \quad a = 95$$

Exercise 16.1

Copy and complete these examples. Check your answers.

	a	b
1	$g + 3 = 9$	$h + 5 = 11$
	$\Rightarrow g = 9 - 3$	$\Rightarrow h =$
	$\Rightarrow g =$	$\Rightarrow h =$
	c	d
	$b - 6 = 8$	$y - 3 = 7$
	$\Rightarrow b = 8 + 6$	$\Rightarrow y =$
	$\Rightarrow b =$	$\Rightarrow y =$

Solve these equations and check your answers.

	a	b
2	$a + 5 = 11$	$b + 8 = 12$
3	$c + 12 = 25$	$x + 9 = 13$
4	$d + 46 = 57$	$y + 14 = 27$
5	$e + 47 = 59$	$s + 45 = 78$
6	$e + 49 = 98$	$d + 97 = 105$
7	$d + 2 \cdot 3 = 7 \cdot 9$	$f + 5 \cdot 8 = 7 \cdot 6$
8	$8 + f = 32$	$9 + g = 64$
	c	d
2	$d - 6 = 10$	$x - 5 = 11$
3	$y - 6 = 15$	$f - 8 = 20$
4	$p - 13 = 56$	$g - 24 = 39$
5	$h - 26 = 56$	$s - 57 = 92$
6	$f - 97 = 106$	$g - 234 = 456$
7	$h - 3 \cdot 2 = 6 \cdot 9$	$s - 3 \cdot 2 = 5$
8	$d + 2\frac{1}{2} = 5\frac{3}{4}$	$4\frac{2}{3} + w = 7\frac{1}{3}$

Read this before doing the next examples.

$$9 - x = 7$$
$$\Rightarrow 9 = 7 + x$$
$$\Rightarrow 9 - 7 = x$$
$$\Rightarrow x = 2$$

Solve these equations and check your answers.

9 $8-r=8$ $24-g=9$ $25-g=17$ $37-x=9$
10 $13-s=8$ $18-k=11$ $15-r=10$ $17-d=2$

Equations involving multiplying and dividing

Example

Solve the equation $4x = 12$.
You can see that $x = 3$ and this is got by dividing 12 by 4. So we can write

$$4x = 12$$
$$\Rightarrow x = \frac{12}{4}$$
$$\Rightarrow x = 3$$

Example

Solve the equation $9x = 45$
We can say

$$\Rightarrow x = \frac{45}{9}$$
$$\Rightarrow x = 5$$

Example

Solve the equation $\frac{a}{2} = 6$

You can see that $a = 12$ and that this is got by multiplying 6 by 2. So we can write

$$\frac{a}{2} = 6$$
$$\Rightarrow a = 6 \times 2$$
$$\Rightarrow a = 12$$

Example

Solve the equation $\frac{a}{3} = 11$
We can say

$$\Rightarrow a = 11 \times 3$$
$$\Rightarrow a = 33$$

If you look at the last four examples you will see that if a number is moved from one side of the = sign to the other we have to change × into ÷, and ÷ into ×. Again you can remember this as:

change the side
change the sign

You also need to remember that ÷ may be written in more than one way. For example we can write either $12 \div 3 = 4$ or $\frac{12}{3} = 4$.

Example

Solve the equation $5x = 17$ and check the answer.
$$5x = 17$$
$$\Rightarrow x = \frac{17}{5}$$
$$\Rightarrow x = 3\tfrac{2}{5} \quad \text{or} \quad 3.4$$

$$5 \overline{)17.0} \quad \frac{3.4}{} \div$$

Check $5x = 5 \times 3.4 = 17.0$.

Example

Solve the equation $\frac{y}{4} = 2.5$ and check your answer.
$$\Rightarrow y = 2.5 \times 4$$
$$\Rightarrow y = 10$$
Check $\frac{y}{4} = \frac{10}{4} = 2\tfrac{2}{4} = 2\tfrac{1}{2} = 2.5$.

Exercise 16.2

Copy and complete these examples. Check your answers.

	a	b	c	d
1	$3b = 12$	$5b = 15$	$\frac{g}{4} = 3$	$\frac{k}{7} = 3$
	$\Rightarrow b = \frac{12}{3}$	$\Rightarrow b =$	$\Rightarrow g = 4 \times 3$	$\Rightarrow k =$
	$\Rightarrow b =$	$\Rightarrow b =$	$\Rightarrow g =$	$\Rightarrow k =$

Solve these equations and check your answers.

	a	b	c	d
2	$3a = 9$	$6b = 12$	$\frac{a}{3} = 2$	$\frac{v}{4} = 3$
3	$5b = 45$	$7c = 21$	$\frac{b}{5} = 4$	$\frac{w}{7} = 5$
4	$7f = 63$	$9s = 72$	$\frac{c}{6} = 5$	$\frac{x}{5} = 6$
5	$8d = 80$	$6h = 54$	$\frac{d}{8} = 7$	$\frac{y}{2} = 8$
6	$9r = 27$	$8h = 48$	$\frac{e}{9} = 7$	$\frac{z}{8} = 9$
7	$4x = 248$	$5x = 545$	$\frac{h}{6} = 5.3$	$\frac{g}{5.2} = 9$
8	$4v = 23$	$8f = 27$	$\frac{g}{4.3} = 2$	$\frac{h}{0.8} = 5$
9	$2c = 4.8$	$8b = 7.2$	$\frac{u}{6} = 7.1$	$\frac{g}{7} = 0.4$
10	$3y = 4.7$	$7c = 2.6$	$\frac{f}{5} = 3.4$	$\frac{k}{0.7} = 1.7$
11	$0.3m = 9.9$	$0.5w = 3.2$	$\frac{u}{1.6} = 3.8$	$\frac{e}{0.7} = 0.3$

Some other kinds of equations

Here are some examples of how to solve other kinds of equations. Read them and then do the next exercise.

Example

$$4x + 7 = 23$$
$$\Rightarrow 4x = 23 - 7$$
$$\Rightarrow 4x = 16$$
$$\Rightarrow x = \frac{16}{4}$$
$$\Rightarrow x = 4$$

$$5b - 4 = 16$$
$$\Rightarrow 5b = 16 + 4$$
$$\Rightarrow 5b = 20$$
$$\Rightarrow b = \frac{20}{5}$$
$$\Rightarrow b = 4$$

$$\frac{h}{5} - 8 = 3$$
$$\Rightarrow \frac{h}{5} = 3 + 8$$
$$\Rightarrow \frac{h}{5} = 11$$
$$\Rightarrow h = 11 \times 5$$
$$\Rightarrow h = 55$$

$$\frac{t}{8} + 3 = 5$$
$$\Rightarrow \frac{t}{8} = 5 - 3$$
$$\Rightarrow \frac{t}{8} = 2$$
$$\Rightarrow t = 2 \times 8$$
$$\Rightarrow t = 16$$

When you do the next exercise you should check your answers.
For example, if we put $t = 16$ in the last example we get $\frac{16}{8} + 3 = 2 + 3 = 5$ which is correct.

Exercise 16.3

Copy and complete these examples. Check your answers.

1 a
$$3x + 7 = 19$$
$$\Rightarrow 3x = 19 - 7$$
$$\Rightarrow 3x = 12$$
$$\Rightarrow x = \frac{12}{3}$$
$$\Rightarrow x =$$

b
$$5b - 5 = 20$$
$$\Rightarrow 5b = 20 + 5$$
$$\Rightarrow 5b =$$
$$\Rightarrow b =$$
$$\Rightarrow b =$$

c
$$\frac{y}{2} + 5 = 8$$
$$\Rightarrow \frac{y}{2} = 8 - 5$$
$$\Rightarrow \frac{y}{2} = 3$$
$$\Rightarrow y = 2 \times 3$$
$$\Rightarrow y =$$

d
$$\frac{h}{6} - 3 = 5$$
$$\Rightarrow \frac{h}{6} = 5 + 3$$
$$\Rightarrow \frac{h}{6} =$$
$$\Rightarrow h =$$
$$\Rightarrow h =$$

unit 16 / page 115

Solve these equations and check your answers.

	a	b	c	d
2	$3s+5=17$	$3z-7=8$	$\frac{y}{3}+2=4$	$\frac{z}{2}-3=4$
3	$6x+3=27$	$5v-7=23$	$\frac{x}{5}+3=4$	$\frac{y}{3}-2=5$
4	$3z+7=13$	$8a-3=37$	$\frac{e}{4}+5=6$	$\frac{f}{6}-4=5$
5	$8d+2=26$	$6f-6=30$	$\frac{w}{7}+3=3$	$\frac{v}{3}-7=2$
6	$3d+9=9$	$7r-6=1$	$\frac{h}{9}+7=9$	$\frac{k}{2}-1=1$
7	$4q+4=14$	$2v-3=4$	$\frac{g}{2}+5=7$	$\frac{h}{5}-15=17$
8	$3a+25=16$	$23+2a=13$	$\frac{d}{7}+6=2$	$\frac{c}{9}-21=31$
9	$7w+18=14$	$37-2c=46$	$\frac{g}{2}+24=25$	$\frac{g}{2}-24=25$
10	$6b-11=15$	$4r+8=17$	$\frac{k}{9}+56=34$	$\frac{k}{13}-9=19$

Using equations to solve problems

An architect wants a room in a house to have an area of 20 square metres with a length of 5 metres. What is the width?

To solve a problem like this using equations we first need to write down the formula.

Area = Length × Width

Then we put in the numbers that we are given.

$\Rightarrow 20 = 5 \times Width$

We then write this equation in a shorter form.

$\Rightarrow 5W = 20$

This equation is then solved.

$\Rightarrow W = \frac{20}{5}$

$\Rightarrow W = 4$ metres.

We can see that this is correct since $5 \times 4 = 20$.

This example is so easy that most of you could have done it in your heads without using equations. Suppose however he wanted the area to be 21 square metres. The equation would have looked like this:

$5W = 21$

$\Rightarrow W = \dfrac{21}{5}$

$\Rightarrow W = 4 \cdot 2$ metres

```
    4·2
5 ) 21·0
    20 ↓
    10
```

Very few people could work this answer out in their heads.

Exercise 16.4

In each of these examples you must first write down an equation and then solve it. **Check your answers.**

1 What is the width of a room with an area of 48 square metres and a length of 8 metres?

2 What is the width of a room with an area of 24 square metres and a length of 6 metres?

3 What is the width of a room with an area of 18 square metres and a length of 9 metres?

4 What is the width of a room with an area of 32 square metres and a length of 8 metres?

5 What is the width of a room with an area of 21 square metres and a length of 7 metres?

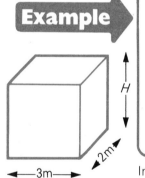

Example

A water tank has to be built on a rectangular base 3 metres by 2 metres. It has to hold 24 cubic metres. What must its height be?

$Volume = Length \times Width \times Height$

$\Rightarrow 24 = 3 \times 2 \times Height$

$\Rightarrow 6H = 24$

$\Rightarrow H = \dfrac{24}{6}$

$\Rightarrow H = 4$ metres

In each of these examples you must first write down an equation and then solve it.
Check your answers.

6 Find the height of a water tank with a volume of 12 cubic metres, a length of 3 metres, and a width of 2 metres.

7 Find the height of a water tank with a volume of 16 cubic metres, a length of 2 metres, and a width of 4 metres.

8 Find the height of a water tank with a volume of 18 cubic metres, a length of 3 metres, and a width of 2 metres.

9 Find the height of a water tank with a volume of 24 cubic metres, a length of 4 metres, and a width of 2 metres.

10 Find the height of a water tank with a volume of 40 cubic metres, a length of 5 metres, and a width of 2 metres.

> **Example**
>
> A sweet pastry is made of one part sugar, two parts butter, and five parts flour. How much of each will be needed to make 16 kg of pastry?
>
> Let the weight of sugar $= w$
> The weight of butter $= 2w$ (Two parts)
> The weight of flour $= 5w$ (Five parts)
>
> The three weights must add up to 16, so we can say:
> $$w + 2w + 5w = 16$$
> $$\Rightarrow 8w = 16$$
> $$\Rightarrow w = 2$$
> Weight of sugar $= 2$ kg
> Weight of butter $= 4$ kg (Two parts)
> Weight of flour $= 10$ kg (Five parts)
> Check: 2 kg $+$ 4 kg $+$ 10 kg $=$ 16 kg.

Find the weights of sugar, butter, and flour needed to make the following weights of pastry.

11 24 kg pastry: One part sugar, two parts butter, five parts flour.

12 14 kg pastry: One part sugar, two parts butter, four parts flour.

13 18 kg pastry: One part sugar, three parts butter, five parts flour.

14 400 g pastry: Two parts sugar, three parts butter, five parts flour.

15 450 g pastry. Two parts sugar, three parts butter, four parts flour.

16 Repeat questions 1 to 5 using the following lengths.
(1) 7 m (2) 5 m (3) 7 m (4) 3 m (5) 9 m.

17 Repeat questions 6 to 10 using the following volumes.
(6) 14 cubic metres (7) 10 cubic metres
(8) 20 cubic metres (9) 25 cubic metres
(10) 42 cubic metres.

18 Repeat questions 11 to 15 using the following weights.
(11) 700 g (12) 800 g (13) 10 kg
(14) 350 g (15) 500 g.

More difficult examples

Example

A number is multiplied by three, seven is added, the answer is 22. What was the number?
Call the number N.
Three times the number is $3N$.
Add seven. $\qquad 3N+7$
This equals 22 so we get the equation
$$3N+7=22$$
$$\Rightarrow 3N = 22-7$$
$$\Rightarrow 3N = 15$$
$$\Rightarrow N = 5$$
The number is 5

Example

The time to cook a joint weighing W kg is given by the formula;
$$T = 30 + 20W \text{ minutes}$$
What is the biggest joint that can be cooked in 2 hrs 10 minutes?
We know that $T=130$ minutes so we can write
$$30 + 20W = 130$$
$$\Rightarrow 20W = 130 - 30$$
$$\Rightarrow 20W = 100$$
$$\Rightarrow W = 5$$
The joint can weigh 5 kg

Example

The cost of a ticket for A adults and C children is given (in pence) by
$$P = 10A + 5C$$
If a total of 140 pence can be spent and 20 children go, how many adults can go?
We know that $P=140$ and $C=20$, so we can write
$$140 = 10A + 5 \times 20$$
$$\Rightarrow 140 = 10A + 100$$
$$\Rightarrow 140 - 100 = 10A$$
$$\Rightarrow 40 = 10A$$
$$\Rightarrow A = 4$$
Four adults can go

Example

A man wishes to save £110 in four years. He decides to save a small amount the first year, an extra £5 the next year, another £5 more the next year, and another £5 more the last year. How much does he need to save the first year?

Let him save during the first year	£s
The second year he will save	£s + 5
The third year he will save	£s + 10
The last year he will save	£s + 15

If we add these up we can see that he will save £4s + 30
But we know that he has to save £110, so we can write

$$4s + 30 = £110$$
$$\Rightarrow 4s = £110 - 30$$
$$\Rightarrow 4s = £80$$
$$\Rightarrow s = £20$$

He has to save £20 the first year

Exercise 16.5

In each of these examples you will need to write down an equation and then solve it. Check your answers.

1. What is the length of a room with an area of 125 square metres and a width of 9 m?
2. What is the length of a room with an area of 91 square metres and a width of 6 m?
3. A number is multiplied by three and five is added to give a total of twenty-three. What is the number?
4. A number is multiplied by seven and ten is subtracted to give an answer of twenty-five. What is the number?

The time to cook a joint weighing W kg is given by the formula $T = 30 + 20W$ minutes.

5. What is the biggest joint that can be cooked in 90 minutes?
6. What is the biggest joint that can be cooked in $2\frac{1}{2}$ hours?

The equation connecting temperatures measured in Centigrade and Fahrenheit is

$$5F = 9C + 160$$

where F is the temperature measured in degrees Fahrenheit, and C is the temperature measured in degrees Centigrade.

7. Change 10 degrees Centigrade to Fahrenheit.
8. Change 20 degrees Centigrade to Fahrenheit.
9. Change 41 degrees Fahrenheit to Centigrade.
10. Change 23 degrees Fahrenheit to Centigrade.

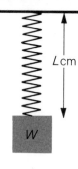

When a weight of W kg is placed on the end of this spring the length in cm is given by $L = 5W + 30$

11 What weight is on the end when the length is 55 cm?

12 What weight is on the end when the length is 45 cm?

13 What weight is on the end when the length is 65 cm?

14 Three men, Smith, Brown, and Jones have to share out profits of £660 from a shop they own. Brown is to get twice as much as Smith, and Jones is to get three times as much as Smith. How much each will they get? (Let S stand for the amount that Smith gets.)

15 Two men run a pop festival and they agree to share any loss or profit so that the second man will get, or have to pay, three times as much as the first man. If the festival makes a loss of £480 how much will each man have to pay? (Let P stand for the amount that the first man will have to pay.)

16 An alloy is made of copper, tin, and zinc. There is twice as much tin as copper, and four times as much zinc as copper. If 140 kg of the alloy is required how much copper, tin, and zinc is needed? (Let C stand for the weight of copper needed.)

17 A sweet pastry is made of one part sugar, two parts butter, and five parts flour. If 400 g of pastry are to be made, what weight of sugar, butter, and flour will be needed?

18 A cement is made from two parts of cement powder, three parts sand, and four parts gravel. If 1800 kg of cement is to be made, what weight of powder, sand, and gravel will be needed? (Let the weight of powder be $2w$, then the weight of sand will be $3w$, etc.)

A firm of printers charges for posters according to the following formula.

$$C = 75 + 2N$$

where N is the number of posters printed, and C is the cost in pence.

19 How many posters can be bought for 97 pence?

20 How many posters can be bought for 155 pence?

21 How many posters can be bought for £2·51?

22 What is the width of a room with a length of 7 m, and an area of 92·4 sq m?

23 A number is multiplied by 7 and 2·5 is added to give an answer of 20. What was the number?

24 In 40 years time I will be three times as old as I am now. How old am I now? (Let my present age be a.)

unit 17 Sharing Things Out

John and Peter do some gardening for John's father. Peter works for twice as long as John. If John's father gave them £6 how should they share it out?

Since Peter works for twice as long as John he should get £2 when John gets £1. This only uses £3, so if we now give Peter another £2 and John another £1, we now use up all of the £6.

Peter gets £2 + £2 = £4   John gets £1 + £1 = £2

We have divided up £6 in the ratio 2 to 1. Another way of writing this is 2:1.

Example

If John and Peter do some more gardening, and John works for 3 days and Peter works for 2 days, how should they share out the £15 that they are paid?

What we have to do is to divide £15 in the ratio 3:2.

First we give John £3 and Peter £2.
We have shared out £5.

Then we give John another £3 and Peter £2.
We have shared out £10.

Then we give John another £3 and Peter £2.
We have shared out £15.

John gets £3 + £3 + £3 = £9
Peter gets £2 + £2 + £2 = £6

We can check that this is probably right since £9 + £6 = £15.

Example

Divide £20 in the ratio 4:1.

If we give out £4 and £1 we have shared out £5.

If we give out another £4 and £1 we have shared out £10.

If we give out another £4 and £1 we have shared out £15.

If we give out another £4 and £1 we have shared out £20.

£4 + £4 + £4 + £4 = £16
£1 + £1 + £1 + £1 = £4

We can check that these are right since £16 + £4 = £20.

Exercise 17.1

Divide these amounts of money in the ratio shown. (Check your answers by the method given above.)

	a	b	c	d	e
1	£9 2:1	£8 3:1	£12 1:2	£12 3:1	£10 2:3
2	£20 3:2	£6 1:1	£15 4:1	£14 3:4	£10 1:4

3 £21 4:3   £14 2:5   £16 5:3   £18 4:5   £24 3:5
4 £21 5:2   £27 5:4   £12 5:1   £14 6:1   £33 6:5
5 £77 5:6   £100 7:3   £91 8:5   £200 9:11   £76 21:17
6 Two men run a shop and agree to share profits in the ratio 2:3.
   a How should they share out profits of £5?
   b How should they share out profits of £15?
   c How should they share out profits of £25?
   d How should they share out profits of £75?
7 Two men paint a house. Fred spends 3 hours, and Jack spends 7 hours.
   a If they are paid £10, how should they share this out?
   b If they are paid £20, how should they share this out?
   c If they are paid £30, how should they share this out?
8 Two men start a business. They agree to share the profits or loss in the ratio 5:3.
   a How should they share a loss of £8?
   b How should they share a loss of £24?
   c How should they share a loss of £32?
   d How should they share a loss of £48?
9 A certain type of concrete is made of powder and sand in the ratio 2:5.
   a What weight of each would be needed to make 28 kg of dry mix?
   b What weight of each would be needed to make 35 kg of dry mix?
   c What weight of each would be needed to make 70 kg of dry mix?
   (Dry mix is the mixture of sand and cement powder before the water is added.)
10 A cake needs flour and sugar in the ratio 4:1.
   a How much of each would be needed to make 5 kg of dry mix?
   b How much of each would be needed to make 10 kg of dry mix?
   c How much of each would be needed to make 20 kg of dry mix?
   d How much of each would be needed to make 40 kg of dry mix?
   (Dry mix is the mixture of flour and sugar before water, milk, eggs, butter, etc., are added.)

More difficult examples

Divide £11 in the ratio 4:3 by the method we have just been using.
You will find that it does not come out exactly. In cases like this it is better to use equations.
First we will use equations to solve a problem that does come out exactly.

Example

> Divide £15 between John and Peter in the ratio 3:2.
>
> You will see that John gets 3 shares and Peter gets 2 shares.
>
> If we let one share be £s we can see that John gets £$3s$ and Peter gets £$2s$.
>
> But we know that they get £15, so we can say:
>
> $$3s + 2s = 15$$
> $$\Rightarrow 5s = 15$$
> $$\Rightarrow s = 15 \div 5$$
> $$\Rightarrow s = 3$$
>
> John gets $3 \times £3 = £9$
> Peter gets $2 \times £3 = £6$
>
> We can check that this is probably right since £9 + £6 = £15.

We will now use equations to solve a problem that does not come out exactly.

Example

> Divide £11 in the ratio 4:3.
> Let one share be £x, so we can say
>
> $$4x + 3x = 11$$
> $$\Rightarrow 7x = 11$$
> $$\Rightarrow x = 11 \div 7$$
> $$\Rightarrow x = 1 \cdot 57$$
>
> One share is £1·57.
> 4 shares = 4 × £1·57 = £6·28
> 3 shares = 3 × £1·57 = £4·71
> Add them to check   £10·99
>
> ```
>        1·57
>     7 )11·00
>        7
>        ---
>        4 0
>        3 5
>        ---
>          50
>          49
>          ---
>           1
> ```

You will see that this does not quite come to £11. This is because when we divided by 7 we did not get an exact answer.

Exercise 17.2

Check your answers by the method shown.
Do some of the examples in Exercise 17.1 using equations, and then do the examples on p. 124.

Divide these amounts of money in the ratio shown.

	a	b	c	d	e
1 £16	3:1	2:3	7:2	4:5	6:5
2 £23	3:2	5:3	3:3	2:5	6:5
3 £46	4:3	2:5	5:3	4:5	3:5
4 £23·57	5:2	5:4	5:1	6:1	6:5
5 £14·93	5:6	7:3	8:5	9:11	21:17

6 Two men run a shop and agree to share profits in the ratio 4:7. How would they share out profits of:

    a £10 **b** £5 **c** £25·60 **d** £3·27?

7 A certain type of concrete is made of cement powder and sand in the ratio 2:5. What weight of each would be needed to make the following weights of dry mix (before water was added):

    a 10 kg **b** 24 kg **c** 32 kg **d** 75 kg?

(Do the divisions to three figures.)

8 A cake needs sugar and flour in the ratio 1:6. How much of each would be needed to make the following weights of dry mix (before water, eggs, butter, etc. were added):

    a 10 kg **b** 6 kg **c** 2 kg **d** 500 g?

(Do the divisions to three figures.)

Making ratios simpler

If we divide £12 in the ratio 1:2 we get £4 and £8.
If we divide £12 in the ratio 2:4 we get £4 and £8.
If we divide £12 in the ratio 4:8 we get £4 and £8.

You should see that these three ratios are the same.

    a 1:2 **b** 2:4 **c** 4:8

If we multiply each number in **a** by 2 we get **b**.
If we multiply each number in **a** by 4 we get **c**.
If we divide each number in **b** by 2 we get **a**.
If we multiply each number in **b** by 2 we get **c**.
If we divide each number in **c** by 2 we get **b**.
If we divide each number in **c** by 4 we get **a**.

If you do the same sort of thing with other ratios you will always find that a ratio is not changed if each of the numbers in the ratio is multiplied or divided by the same number.

Example

Write these ratios in their simplest form.

    a 12:4 **b** 21:14 **c** $3\frac{1}{2}:10\frac{1}{2}$

a We can divide each number by 4 to get 3:1.
b We can divide each number by 7 to get 3:2.

c We can multiply each number by 2 to get 7:21.
We can now divide each number by 7 to get 1:3.

Example

Two men pay 15p and 25p on a football pools coupon. If the coupon wins £32 how much each will they get?

The winnings must be paid out in the ratio 15:25.
By dividing each number by 5 we can write the ratio 3:5.
Let one share be £y, and we can write:

$$3y + 5y = 32$$
$$\Rightarrow 8y = 32$$
$$\Rightarrow y = 32 \div 8$$
$$\Rightarrow y = 4$$

Their shares are $3 \times £4 = £12$ and $5 \times £4 = £20$.
Check: £12 + £20 = £32

Exercise 17.3

Write these ratios in their simplest form.

	a	b	c	d	e
1	6:8	4:6	10:4	4:2	6:9
2	9:12	8:12	20:8	12:20	15:9
3	4:8	6:15	7:3	6:10	8:5
4	$2\frac{1}{2}:7\frac{1}{2}$	$5:7\frac{1}{2}$	$5\frac{1}{4}:2\frac{1}{4}$	$\frac{3}{4}:\frac{1}{4}$	$2\frac{2}{3}:\frac{2}{3}$
5	2·1:1·4	2·5:7·5	7·5:30	0·9:0·3	1·5:2·5

6 A man and his wife invested £200 and £300 in a business.
  a Write the ratio 200:300 in its simplest form.
  b How would they share out profits of £50?
  c How would they share out profits of £75?
  d How would they share out profits of £32?

7 70 kg of alloy contains 40 kg copper and 30 kg zinc.
  a Write the ratio 40:30 in its simplest form.
  b What weights of copper and zinc would be needed for 28 kg of the alloy?
  c What weights of copper and zinc would be needed for 35 kg of the alloy?
  d What weights of copper and zinc would be needed for 85 kg of the alloy?

8 Two brothers share a car. In one month they drove 500 km and 300 km.
  a Write the ratio 500:300 in its simplest form.
  b If the car cost £16 to run in that month, how much should each pay?

c If the car cost £20 to run in that month, how much should each pay?

d If the car cost £23 to run in that month, how much should each pay?

9 Two men win £80 between them on a football pools coupon.

   a How much each will they get if they invested 30p and 50p?

   b How much each will they get if they invested 70p and 30p?

   c How much each will they get if they invested 15p and 25p?

   d How much each will they get if they invested 40p and 30p?

10 Two men invest 10p and 15p on a pools coupon. How much each will they get if the coupon wins:
   a £100 **b** £110 **c** £776 **d** £34567 **e** £346 643?

Ratios with three or more numbers

Ratios with more than two numbers in them are dealt with in the same way as those with two numbers.

Example

Divide £45 in the ratio $3:2:4$.

If we let one share be y, we can write

$$3y + 2y + 4y = 45$$
$$\Rightarrow 9y = 45$$
$$\Rightarrow y = 45 \div 9$$
$$\Rightarrow y = 5$$

$3 \times £5 = £15$; $2 \times £5 = £10$; $4 \times £5 = £20$

Check: $£15 + £10 + £20 = £45$

Example

Write this ratio in its simplest form.

$$12:9:15$$

We can divide each number by 3 and get $4:3:5$.

Example

Three men bet on a horse. They invest 15p, 20p and 10p. Total winnings are 180p, how do they share it?

They have to share the winnings in the ratio $15:20:10$.

We can divide each number by 5 and get $3:4:2$.

If we let one share be t pence we can write:

$$3t + 4t + 2t = 180$$
$$\Rightarrow 9t = 180$$
$$\Rightarrow t = 180 \div 9$$
$$\Rightarrow t = 20$$

They get $3 \times 20p = 60p$ $4 \times 20p = 80p$ $2 \times 20p = 40p$

Exercise 17.4

The first 6 of these examples come out exactly, and they can be done by either method.

Divide these quantities in the ratio shown.

	a		b		c	
1	£6	1:2:3	£9	2:3:4	£11	5:4:2
2	18p	2:3:4	22p	5:4:2	16p	2:3:3
3	33 m	5:4:2	24 m	2:3:3	24 m	1:2:3
4	32 kg	2:3:3	30 kg	1:2:3	45 kg	2:3:4

	d		e	
1	£8	2:3:3	£12	1:2:3
2	18p	1:2:3	27p	2:3:4
3	36 m	2:3:4	44 m	5:4:2
4	55 kg	5:4:2	40 kg	2:3:3

5 Three men run a shop and agree to share profits in the ratio 3:4:1. How should they share out profits of:

a £16   b £24   c £32   d £40   e £200?

6 Three men paint and decorate some rooms in a house. John spends 2 hours, Brian spends 3 hours, Harry spends 5 hours. How much should each get if they are paid:

a £10   b £20   c £30   d £40   e £5?

7 Three women start a business. They agree to share profits or losses in the ratio 1:3:5. How should they share out a loss of:

£8   £16   £24   £40   £200?

Divide these quantities in the ratio shown.

		a	b	c	d	e
8	£25	1:2:3	2:3:4	5:4:2	2:3:3	3:4:1
9	36 kg	2:3:5	1:3:5	3:5:1	4:1:3	2:4:5
10	42 m	4:3:2	4:5:6	6:7:8	9:2:9	7:9:12

11 A certain type of concrete is made of cement powder, sand, and gravel in the ratio 1:2:4. What weight of each would be needed to make the following weights of dry mix:

a 50 kg   b 43 kg   c 10 tonnes   d 5 tonnes
e 1 tonne?

12 A cake needs sugar, flour, and butter in the following ratio 1:4:2. It also needs 1 egg for each 200 g of mixture. What weight of sugar, flour, and butter is needed to make the following weights of mix (before eggs, water, etc. are added)? How many eggs will be needed?

a 200 g   b 400 g   c 800 g   d 1 kg   e 5 kg?
(1 kg = 1000 g).

Write these ratios in their simplest form.

	a	b	c
13	2:4:6	6:8:2	4:6:8
14	15:12:6	5:10:15	8:12:16
15	$2\frac{1}{2}:3\frac{1}{2}:4\frac{1}{2}$	$2\frac{1}{2}:5:7\frac{1}{2}$	1·2:1·5:1·8

	d	e
13	10:8:4	14:21:35
14	100:200:100	150:200:300
15	$1\frac{1}{3}:2\frac{2}{3}:4$	0·2:0·3:0·6

16 Three men invest 10p, 20p, 30p in a pools coupon.
   a Write the ratio 10:20:30 in its simplest form.
   b How should they share out winnings of £12?
   c How should they share out winnings of £20?
   d How should they share out winnings of £100?
   e How should they share out winnings of £200 000?

17 Three men win £78 between them on a pools coupon.
   a How much will each get if they invested 30p, 20p, and 40p?
   b How much will each get if they invested 12p, 15p, and 18p?
   c How much will each get if they invested 50p, £1, and £1·50?
   d How much will each get if they invested 75p, £1·25, and £2?

18 100 kg of an alloy contains 50 kg copper, 30 kg tin, and 20 kg zinc.
   a Write the ratio 50:30:20 in its simplest form.
   b What weight of each metal would be needed to make 180 kg of the alloy?
   c What weight of each metal would be needed to make 200 kg of the alloy?
   d What weight of each metal would be needed to make 50 g of the alloy?
   e What weight of each metal would be needed to make 250 tonnes of the alloy?

19 Three families hire a holiday cottage for 9 weeks. During this time the families use it for 28 days, 14 days, and 21 days.
   a Write the ratio 28:14:21 in its simplest form.
   b If the total cost of hiring the cottage is £100, how much will each family have to pay?
   c If the total cost of hiring the cottage is £155, how much will each family have to pay?
   d If the total cost of hiring the cottage is £123·89, how much will each family have to pay?

unit 18 The Circle

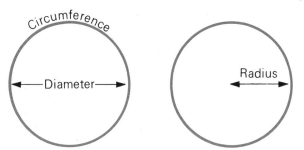

The distance across a circle is called the **diameter**.
The distance half way across a circle is called the **radius**.
The distance around a circle is called the **circumference**.

What is the diameter of this circle?
What is the radius of this circle?

What is the radius of this circle?
What is the diameter of this circle?

You should see that the diameter of a circle is twice the radius.

Finding the distance around a circle

Draw some circles on thick cardboard with diameters as shown below; find the distances around the circles by rolling them along the edge of a ruler or winding cotton around each one and then measuring the length of the cotton. Copy the table below and fill in your answers. Some of them have been done for you.

Diameter in cm	1	2	3	4	5	6	7	8	9	10
Circum- ference in cm	3·2		9·4			18·8				31·2

You can see that the distance around each circle is just over three times the distance across.

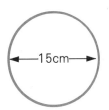

The distance around this circle = 3 × 15 = 45 cm.

This is only a rough answer and a more exact value can be found by using 3·1 instead of 3. A more exact value still can be found using 3·14. This number is used so often in mathematics that the Greek letter π is used to stand for it. This letter is called pi.

The important thing to remember is that

circumference = π × diameter

You can also write it in a shorter way like this.

C = πD

Example

Find the distance around this circle
Circumference = 3·1 × 7 = 21·7 mm

Exercise 18.1

In this exercise you can use either 3, or 3·1, or 3·14 for π. Your teacher will tell you which value to use.

Find the distances around these circles.

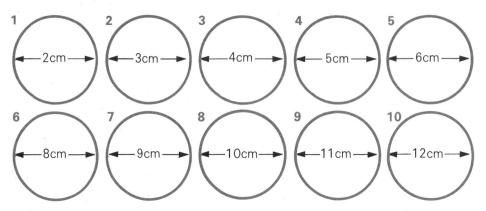

11 What is the diameter of this circle?
    What is the distance around this circle?

Find the distances around these circles.

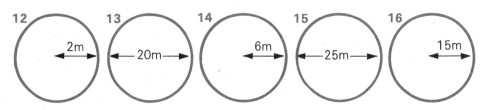

Finding the area of a circle

Make two copies of the diagram shown below on thick paper or cardboard. From one copy of the diagram cut out a circle, and from the other one cut out a square. You should now have a circle that fits into the square. Cut the square into four smaller squares and see how many of them balance the circle when they are placed on a pair of scales. You should find that the circle is slightly heavier than three of the squares. In other words we can say that approximately

area of circle = 3 × area of small square

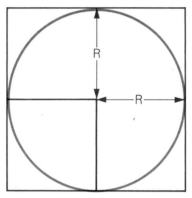

Since the area of a small square = **radius × radius**

we can write **area of circle = 3 × radius × radius**

It can be shown that to get the exact answer we have to use π instead of 3. So we can write

Area of circle = π × radius × radius

This may also be written

$$A = \pi R^2$$

Example

Find the area of this circle.
Area = 3·1 × 5 × 5 = 77·5 sq cm

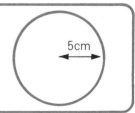

Since π = 3·142 to three decimal places and $\frac{22}{7}$ = 3·143 to three decimal places $\frac{22}{7}$ can be used instead of π. This is particularly useful if the 7 can cancel out with another number.

page 132/unit 18

Example

Find the area of this circle.
Area = $\frac{22}{7} \times 14 \times 14 = 22 \times 2 \times 14$
     = 616 sq cm

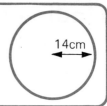

Exercise 18.2

In this exercise you can use either 3, or 3·1, or 3·14 for π. Your teacher will tell you which value to use.

Find the areas of these circles.

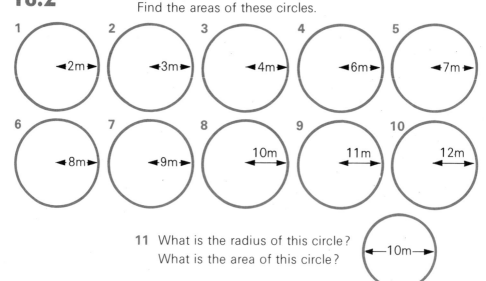

1. 2m
2. 3m
3. 4m
4. 6m
5. 7m
6. 8m
7. 9m
8. 10m
9. 11m
10. 12m

11 What is the radius of this circle?
    What is the area of this circle? — 10m —

Find the areas of these circles.

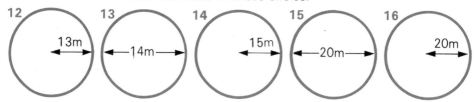

12. 13m
13. 14m
14. 15m
15. 20m
16. 20m

Some practical problems

There are many practical problems where the area and circumference need to be found.

Example

In an experiment 25 turns of copper wire have to be wrapped around a wooden rod which has a diameter of 6 cm. How much wire will be needed?

Distance around rod = 3·1 × Diameter = 3·1 × 6
                                     = 18·6 cm

Length of wire = 25 × 18·6 cm = 465 cm

Example

A circular roof with a radius of 9 m has to be painted. One litre of paint is needed for every 8 square metres. How many litres will be needed to paint the roof?

Area = 3·1 × 9 × 9 = 251·1 square metres.
Number of litres required = 251·1 ÷ 8 = 31·3 litres

Exercise 18.3

In this exercise you can use either 3, or 3·1, or 3·14 for π. Your teacher will tell you which value to use.

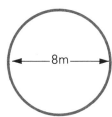

1. A man wishes to put a metal edging around the edge of the lawn shown.
   a Find the distance around the lawn.
   b How much will the strip cost if one metre cost 6p?

2. A circular roof with a radius of 3 m has to be painted. Each square metre will cost 7p to paint.
   a Find the area of the roof.
   b What is the cost of painting the roof?

3. A shopkeeper who sells plastic piping, notices that he has seven turns left on the roll which has a diameter of 2 m.
   a What is the distance around the roll?
   b What length of plastic piping is there left?

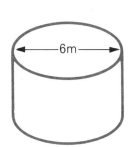

4. The top of the storage tank shown here is made of sheet steel which weighs 9 kg per square metre.
   a What is the radius of the top of the tank?
   b What is the area of the top of the tank?
   c What is the weight of metal used in the top of the tank?

5. The tank shown above has a safety rail fixed around the top.
   a What length of safety rail will be required?
   b What will be the cost of the safety rail if one metre costs £3?

6. A firm which lays dance floors charges £50 plus £5 for each square metre. If a circular dance floor is required with a diameter of 18 m.
   a What is the area of the dance floor?
   b What is the cost of laying the floor?

7. A circular floor of 3 m radius is to be covered with small mosaic tiles. 2500 tiles are needed for each square metre. How many tiles will be needed to cover the floor?

8 A circular flower bed is to be surrounded by a fence with three strands of wire. The posts are to be 2 m apart. The radius of the flower bed is 5 m.

  a What is the distance around the flower bed?

  b How many posts will be needed?

  c What is the total length of wire needed?

9 A conveyor belt is driven by a wheel which turns twice every second. The radius of the wheel is 1·5 m.

  a Find the circumference of the wheel.

  b How far will the conveyor belt move every second?

  c How far will the conveyor belt move every minute?

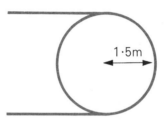

10 A cake shop has to supply an order for 12 cakes, the top of which are covered with a coffee flavoured chocolate which will have to be specially ordered. The cakes have a diameter of 20 cm, and the chocolate is supplied in blocks, each one of which covers about 400 sq. cm.

  a What is the area of the top of each cake?

  b What is the total area of all the cakes?

  c How many blocks of chocolate will have to be ordered?

11 A woman makes a table set which consists of six table mats with diameter 24 cm, and a table cloth with a diameter of 1·7 m. She decides to edge them all with lace which costs £1·73 per metre.

  a What is the total length of lace required?

  b How much will it cost?

unit 19 What are the Chances?

There are five pupils in a cookery class. Jane, Mark, Jill, Tim, and Susan. Two of them are to be chosen to cook a meal and serve it to some visitors to the school. Jane wonders what the chances are of her and her friend Mark being chosen.

One way of solving this problem is to write down all the possible pairs.

Jane	Jane	Jane	Jane	Mark	Mark	Mark	Jill	Jill	Tim
Mark	Jill	Tim	Susan	Jill	Tim	Susan	Tim	Susan	Susan

You will see that there are ten possible pairs of pupils, so we say that the chance of Jane and Mark being chosen is 1 in 10. Another name for chance is **probability**, and we often write the probability as a fraction. So we can say:

probability of Jane and Mark being chosen $=\frac{1}{10}$

Example

What is the chance of Jane being chosen?

Out of the ten possible pairs you will see that Jane is in four of them so we can say that:

probability of Jane being chosen $=\frac{4}{10}$

Example

What is the chance of Jane not being chosen?

Out of ten possible pairs Jane does not appear in six of them so we can say that:

probability of Jane not being chosen $=\frac{6}{10}$

Exercise 19.1

There are five girls in a cookery class, we will call them A, B, C, D, E. If three of them are chosen there are ten possible groups of three. They are:

ABC, ABD, ABE, ACD, ACE,
ADE, BCD, BCE, BDE, CDE.

Find the following probabilities:

1 That A is chosen.

2 That A and B are chosen.

3 That A and B and C are chosen.

4 That D is chosen.

5 That D and C are chosen.

6 That D and C and E are chosen.

7 That E is not chosen.

8 That B or C (or both of them) are chosen.

9 That either B or C (but not both of them) are chosen.

10 That at least one of A, or B, or C are chosen.

11 a What is the probability that C is chosen?

  b What is the probability that C is not chosen?

c Add the answers to **a** and **b**.

12 a What is the probability that A, C, E are all chosen?

  b What is the probability that none of A, C, E are chosen?

There are four boys in a woodwork group, we will call them P, Q, R, S. Two of them are chosen to take part in the school exhibition.

13 Write down all the possible groups of two. (There are 6 possible pairs.) Find the following probabilities.

14 That P is chosen.

15 That P is not chosen.

16 That P and Q are chosen.

17 That P and Q are not chosen.

18 That R or P (or both of them) are chosen.

19 That either R or P (but not both of them) are chosen.

20 That P or R or S (or any two of these three are chosen).

In the previous work it did not matter which way around we chose the groups of pupils. In some problems it does.

Example

Four cards marked with the numbers 1, 2, 3, 4 are placed in a box.

One of the cards is taken out, and then a second card.

a Write down all the possible pairs that can be taken out. Write down the following probabilities.

b That the number 23 is drawn.

c That the number 41 is drawn.

d That a number between 20 and 30 is drawn.

e That a number between 30 and 40 is drawn.

f That a number less than 50 is drawn.

g That a number less than 10 is drawn.

a 12, 21, 13, 31, 14, 41, 23, 32, 24, 42, 34, 43. (There are twelve.)

b Probability of 23 being drawn $=\frac{1}{12}$.

c Probability of 41 being drawn $=\frac{1}{12}$.

d There are 3 numbers between 20 and 30 so probability $=\frac{3}{12}=\frac{1}{4}$.

e There are 3 numbers between 30 and 40 so probability $=\frac{3}{12}=\frac{1}{4}$.

> f All 12 numbers are less than 50 so
>   probability $=\frac{12}{12}=1$.
> g There are 0 numbers less than 10 so
>   probability $=\frac{0}{12}=0$.

You should see that when something is **certain** to happen, as in **f,** the probability is 1.

When something is **impossible,** as in **g,** the probability is 0.

Exercise 19.2

Three cards marked with the numbers 2, 5, 6 are placed in a box.

1. Write down all the possible pairs that can be taken out. (There are 6.) Write down the following probabilities.
2. That the number 25 is drawn.
3. That the number 65 is drawn.
4. That the number 75 is drawn.
5. That a number between thirty and sixty is drawn.
6. That a number starting with 5 is drawn.
7. That a number ending or starting with 5 is drawn.
8. That a number between 30 and 40 is drawn.
9. That a number between 10 and 70 is drawn.

Three cards marked with the letters T, E, A are placed in a box. One of the cards is taken out, and then a second card and then the third card.

10. Write down all the possible groups of three that can be taken out. Write down the following probabilities.
11. That the three cards make the word TEA.
12. That the three cards make the word ATE.
13. That the three cards make a word starting with T.
14. That the three cards make a word ending with E.
15. That the three cards contain the letter E before the letter A.
16. That the three cards have the letter T in the middle.
17. That the three cards do not have the letter T in the middle.
18. Add together the answers to 16 and 17.

Four cards marked with the letters A, T, E, R are placed in a box. One of the cards is taken out, and then a second card.

19. Write down all the possible pairs that can be taken out.

Write down the following probabilities.

That the two cards make the word AT.

page 138/unit 19

21 That the two cards contain the letter T.
22 That the two cards contain the letters A and T.
23 That the two cards contain either the letter A or E (or both).
24 That the two cards contain either the letter A or E (but not both).
25 That the two cards do not contain the letter A.
26 That the two cards contain neither the letter A nor the letter R.

Using probabilities

In the examples you have just done you worked out the number of times something happened, and then worked out the probability. You are now going to learn to work out the number of times something can happen if you are given the probability.

Example

It is known that in a certain school the probability of a pupil failing a swimming test is 1 in 4 (that is $\frac{1}{4}$). How many will fail if 12 take the test? Since $\frac{1}{4}$ will fail we have to work out $\frac{1}{4}$ of 12.

$$\frac{1}{{}_1\cancel{4}} \times \frac{\cancel{12}^3}{1} = 3 \qquad \text{(4 cancels into the top and bottom)}$$

Example

It is known that the probability of a particular tyre lasting more than 20 000 km is $\frac{2}{5}$. If 5 cars have these tyres fitted, how many will need replacing after 20 000 km?

There are 20 tyres. Since only $\frac{2}{5}$ of the tyres last over 20 000 km we have to work out $\frac{2}{5}$ of 20.

$$\frac{2}{{}_1\cancel{5}} \times \frac{\cancel{20}^4}{1} = 2 \times 4 = 8.$$

This is the number of tyres that last, so the number of tyres that wear out $= 20 - 8 = 12$.

Another way of getting the answer is by working out $\frac{3}{5} \times 20$. Why?

In these examples we are not saying that exactly three pupils will fail the test, or that exactly 12 tyres will be needed. Four pupils could fail, or we might only need ten tyres.

What we are saying is that we are more likely to have 3 pupils failing the test than any other number, and that we are more likely to need 12 new tyres than any other number.

Probabilities are very useful in science, engineering, and many other jobs, and the next exercise will show you how they are used.

Exercise 19.3

1. If the probability of a boy failing a swimming test is $\frac{2}{5}$, how many would you expect to fail if the number that took the test was:

   a 10 boys **b** 20 boys **c** 25 boys **d** 35 boys **e** 60 boys?

2. If the probability of a girl failing a swimming test is $\frac{1}{5}$, how many would you expect to fail if the number that took the test was:

   a 10 girls **b** 20 girls **c** 25 girls **d** 35 girls **e** 60 girls?

3. The probability of a certain type of tyre being worn out after 40 000 km is $\frac{3}{5}$. How many would you expect to be worn out after 40 000 km if there were:

   a 10 tyres **b** 15 tyres **c** 20 tyres **d** 35 tyres **e** 70 tyres?

4. The probability of a certain type of tyre being worn out after 30 000 km is $\frac{4}{5}$. How many would you expect to be worn out after 30 000 km if there were:

   a 10 tyres **b** 20 tyres **c** 25 tyres **d** 40 tyres **e** 65 tyres?

5. A factory uses drilling machines. The probability of a machine breaking down in one year is $\frac{3}{10}$. How many breakdowns can they expect in one year if they have:

   a 20 machines **b** 30 machines **c** 40 machines **d** 60 machines **e** 200 machines?

6. A car-hire firm finds that the probability of one of its cars breaking down in one month is $\frac{2}{9}$. How many breakdowns can they expect each month with:

   a 27 cars **b** 18 cars **c** 36 cars **d** 45 cars **e** 90 cars?

7. A calculating machine firm finds that the probability of one of its machines breaking down in a year is $\frac{1}{7}$. How many machines will it have to repair each year if it has sold:

   a 28 **b** 35 **c** 56 **d** 70 **e** 700

8. A T.V. rental firm finds that the probability of one of its T.V. sets breaking down in one year is $\frac{2}{9}$. How many breakdowns will it have each year if it rents out:

   a 27 sets **b** 72 sets **c** 81 sets **d** 90 sets **e** 306 sets?

In the next problems you have first to work out a probability, and then use the probability to answer some more questions.

Example

> A firm making car batteries tested 20 of them at the end of two years and found that 4 had worn out.
> a What was the probability of a battery wearing out by the end of two years?
> b How many batteries out of 45 would you expect to be worn out by the end of two years?
>
> a Probability of battery wearing out $= \dfrac{4}{20} = \dfrac{1}{5}$
> b Number of batteries wearing out $= \dfrac{1}{5} \times 45 = 9$

9 A firm making car batteries tested 20 of them at the end of two years and found that 8 had worn out.
  a What is the probability of a battery wearing out at the end of two years? At the end of two years how many batteries would you expect to be worn out if there were:
  b 25 batteries   c 45 batteries   d 50 batteries
  e 75 batteries?

10 A factory making chairs found that on average six out of every 24 chairs they made had slight faults and needed extra work done on them before they could be sent to the shops.
  a Find the probability of a chair having faults. How many chairs would you expect to have faults if the factory made:
  b 16 chairs   c 32 chairs   d 48 chairs
  e 40 chairs   f 144 chairs?

11 A car insurance firm found that on average out of each 70 motorists insured with them, 20 made a claim during one year.
  a What is the probability of a motorist making a claim during one year? Find how many claims the firm can expect each year if they insure:
  b 42 motorists   c 49 motorists
  d 56 motorists   e 238 motorists?

12 A firm making light bulbs tested 36 of them for 2000 hours. At the end of 2000 hours 8 had failed.
  a Work out the probability of a light bulb failing during the first 2000 hours. Work out how many light bulbs would be expected to fail during the first 2000 hours if the number of bulbs used was:
  b 18   c 27   d 45   e 72   f 216   g 333.

13 A pottery manufacturer found that out of every 45 pots he produced, 5 had a slight flaw and had to be sold cheaply.

   a What is the probability of a pot having a flaw? How many pots would he expect to sell off cheaply if he produced:
   b 18 pots **c** 36 pots **d** 72 pots **e** 81 pots
   f 486 pots?

Often a probability is given as a decimal instead of as a fraction.

> **Example**
>
> The probability of a light bulb failing during the first 3000 hours is 0·3. How many would you expect to fail during the first 3000 hours if there were:
>
> **a** 20 bulbs **b** 37 bulbs?
>
> **a** Number failing $= 0·3 \times 20 = 6$ bulbs.
> **b** Number failing $= 0·3 \times 37 = 11·1$ bulbs.
>
> Since obviously only a whole number of bulbs can fail we take the nearest whole number 11. If the answer came to 11·8 the answer would be 12.

Do examples 1 to 8 using the following probabilities.
1 0·4 **2** 0·2 **3** 0·6 **4** 0·8 **5** 0·3 **6** 0·22 **7** 0·14 **8** 0·22.

Gambling and probability

Many gambling games involve the probability of certain things happening, like getting two heads when two coins are thrown, or getting four aces out of a pack of cards. We will now deal with the problem of two coins.

If we throw two coins then we can either get 2 heads, or 2 tails, or 1 of each. There are three possible ways that the coins can fall so we can say:

     Probability of 2 heads $= \frac{1}{3}$
     Probability of 2 tails $= \frac{1}{3}$
     Probability of 1 head and 1 tail $= \frac{1}{3}$

If we throw the two coins say 24 times we would expect to get about 8 lots of 2 heads, 8 lots of 2 tails, and 8 lots of 1 head and 1 tail. If we put our results on a bar chart we would expect to get a chart looking something like this with roughly equal numbers of each.

	1	2	3	4	5	6	7	8	9	10
2 heads										
1 of each										
2 tails										

Get two coins, throw them 24 times, and draw a bar chart of your results. Look at the bar charts of your friends. Most of them will have a shape different to the bar chart above. There must be something wrong with the probabilities of $\frac{1}{3}$ we got above. Let us look at the argument more closely.

If we throw two coins we can get either

You can see that there are four possible ways, not three, so we can say:

Probability of 2 heads $=\frac{1}{4}$
Probability of 1 head and 1 tail $=\frac{2}{4}=\frac{1}{2}$
Probability of 2 tails $=\frac{1}{4}$

If we throw the two coins say 24 times we would expect to get about 6 pairs of heads, 12 pairs with 1 head and 1 tail, and 6 pairs of tails. If we put our results on a bar chart we would get a chart looking something like this.

1 2 3 4 5 6 7 8 9 10 11 12 13 14

Look at your bar charts, they will probably be more like this. The probabilities of $\frac{1}{3}$ given first are wrong. The correct probabilities are $\frac{1}{4}$, $\frac{1}{2}$, and $\frac{1}{4}$.

Exercise 19.4

Three coins are thrown. There are 8 ways they can land. Three of the ways are shown below.

1 Copy this and write down the other 5 ways.

2 Copy and complete the following:

   a Probability of 3 heads =—

   b Probability of 2 heads and 1 tail =—

   c Probability of 1 head and 2 tails =—

   d Probability of 0 heads and 3 tails =—

unit 19/page 143

3 Three coins are thrown 8 times; copy and complete this table which shows the number of times you might expect to get 3 heads, 2 heads, etc.

3 heads 0 tails	2 heads 1 tail	1 head 2 tails	0 heads 3 tails
1			

4 Show the results in question **3** on a bar chart.

5 Throw 3 coins 8 times and make a bar chart of your results. Compare your bar chart with the one you got in question **4**.

Four coins are thrown. There are 16 ways they can land. Three of the ways are shown below.

6 Copy this and write down the other 13 ways.

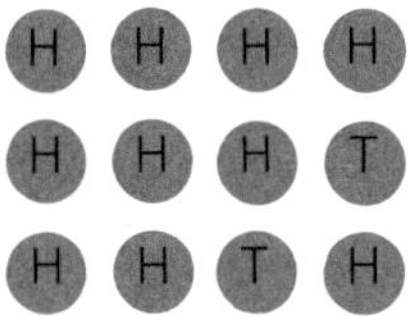

H H H H
H H H T
H H T H

7 Copy and complete the following:
  a Probability of 4 heads = —
  b Probability of 3 heads and 1 tail = —
  c Probability of 2 heads and 2 tails = —
  d Probability of 1 head and 3 tails = —
  e Probability of 0 heads and 4 tails = —

8 Four coins are thrown 32 times; copy and complete this table which shows the number of times you might expect to get 4 heads, 3 heads, etc.

4H 0T	3H 1T	2H 2T	1H 3T	0H 4T
2				

9 Show the results of question **8** on a bar chart.

10 Throw 4 coins 32 times and make a bar chart of your results. Compare your bar chart with the one you got in question 9.

Throwing dice

A single dice was thrown 36 times. The number of times each number came up is shown on bar chart A below. It was then thrown another 36 times and the results of the 72 throws is shown on bar chart B. It was then thrown another 36 times and the result of the 108 throws is shown on bar chart C.

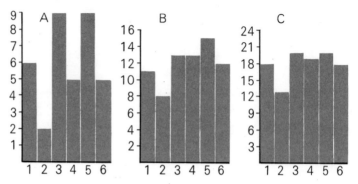

You will see that the first chart is rather bumpy, B is less bumpy, and C is less bumpy still. What sort of bar chart would we expect to get if we threw the dice a very large number of times?

Since there are 6 numbers, and there is no reason why we should get any number more often than any other number, we should expect roughly equal numbers of each. The top of the bar chart should be roughly level.

Throw a dice 36 times or 72 times (or any number that 6 will go into) and draw the bar chart of your results. Collect together the results from the rest of the class and draw a bar chart showing all the results. Is the top of this bar chart more level than yours?

throwing two dice

If two dice are thrown there are eleven possible scores from 2 to 12. If two dice are thrown a large number of times what sort of bar chart should we get? If we throw the dice say 88 times should we expect to get eight 2's, eight 3's, eight 4's and so on, with all the bars on the bar chart being roughly the same height?

A pair of dice was thrown 88 times; the results are shown on the bar chart below.

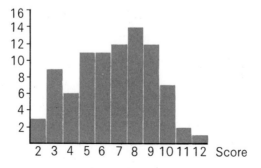

Throw a pair of dice and make a bar chart of your results. Look at the bar charts of your friends. Most of them seem to be higher in the middle. It seems that scores around 7 seem to happen more often than high or low scores. Why is this?

If we write down all the possible ways that 2 dice can land we find that there are 36. They are shown in the table below.

		\multicolumn{6}{c}{Number on second die}				

		1	2	3	4	5	6
Number on first die	1	2	3	4	5	6	7
	2	3	4	5	6	7	8
	3	4	5	6	7	8	9
	4	5	6	7	8	9	10
	5	6	7	8	9	10	11
	6	7	8	9	10	11	12

You can now see that there is only one way that a score of 2 or 12 can be made, but that scores of numbers like 6 and 7 can be made in several ways. For example a score of 7 can be made by 6 on the first dice and 1 on the second, or 5 on the first dice and 2 on the second, and so on.

A bar chart of the possible scores is shown below. How does it compare with your bar chart?

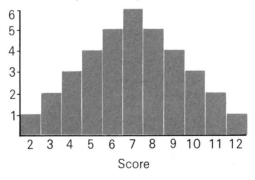

We can use the table which shows all the possible 36 ways that 2 dice can land to do probability problems.

Example

a What is the probability of a score of 5 with two dice?

b What is the probability of a score of less than five with two dice?

c If two dice are thrown 12 times how many times might we expect a score of less than five?

a Number of ways of scoring 5 = 4.
   Probability of scoring 5 = $\frac{4}{36} = \frac{1}{9}$
b Number of ways of scoring less than 5 = 6.
   Probability of scoring less than 5 = $\frac{6}{36} = \frac{1}{6}$
c Using the answer to **b** we get
   Expected number of scores of less than 5 = $\frac{1}{6} \times 12 = 2$

This answer does not mean that if we throw a pair of dice 12 times we will get just 2 scores of less than 5, but that we are more likely to get 2 scores of less than 5 than any other number. If we threw a pair of dice 12 times, and kept on repeating this, we would find that on average we would get 2 scores of less than five each time.

Exercise 19.5

Two dice are thrown. Find the probability of:

1 A score of 7.
2 A score of 11.
3 A score of 12.
4 A score of 8.
5 A score of 6.
6 A score of 13.
7 A score of more than 8.
8 A score of 8 or more.
9 A score of less than 7.
10 A score of 7 or less.
11 A score of 6, 7, or 8.
12 A score of 2, 3, 4, or 5.

Two dice are thrown 72 times. Use the answers you got above to answer the following questions.

In how many of the 72 throws would you expect to get:

13 A score of 7.
14 A score of 11.
15 A score of 12.
16 A score of 8.
17 A score of 6.
18 A score of 13.
19 A score of more than 8.
20 A score of 8 or more.
21 A score of less than 7.
22 A score of 7 or less.
23 A score of 6, 7, or 8.
24 A score of 2, 3, 4, or 5?

Do questions **13** to **24** again for 24 throws. Some of the answers will not come out exactly, and you write them correct to the nearest whole number.

Example

A pair of dice is thrown 48 times. How many times can we expect to get a score of 11?

From question **2** we see that the probability = $\frac{1}{18}$
(Since $\frac{2}{36} = \frac{1}{18}$)
Expected number of scores of 11 = $\frac{1}{18} \times 48 = \frac{8}{3} = 2.66$
3 times

unit 20 Graphs

Understanding graphs

The graph shows how the temperature changed during a day in August. From a graph like this much information can be obtained.

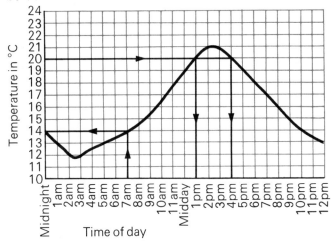

Example

What was the temperature at 7 a.m.?

Find 7 a.m. along the bottom of the graph. Follow the line up until the curved graph line is reached. Read along the straight line to the side of the graph. You will see that the temperature at 7 a.m. was 14 °C.

Example

At what time of day was the temperature 20 °C?

Find 20 °C at the side of the graph. Follow the line along until the curved graph line is reached. Read down the straight line to the bottom of the graph. In this case you will see that there are two answers. 1 p.m. and 4 p.m.

Use the graph to see if you can answer these questions:

  a What was the temperature at 8 p.m.?
  b What was the temperature at 11 a.m.?
  c At what times was the temperature 19 °C?
  d At what times was the temperature 13 °C?
  e What was the highest temperature during the day?
  f What was the lowest temperature during the day?
  g At what time was the temperature at its highest?
  h At what time was the temperature at its lowest?

The lines with the numbers at the bottom and side of the graph are called the **axes**.
The one along the bottom is called the **horizontal axis**, and the one at the side is called the **vertical axis**.
When graphs are drawn they must always have a heading to explain what they are about, and the axes must have labels on them.

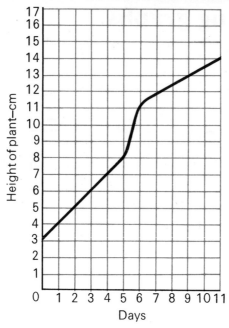

Exercise 20.1

1. The graph shows the height of a plant cutting for the first 11 days after planting. Use the graph to answer the following questions.

   a How tall was the cutting on planting?
   b How much did it grow during the first day?
   c After how many days has the cutting grown 3 cm?
   d During which day does the cutting grow most?
   e Which days does it grow least?
   f How tall is the cutting after 10 days?
   g How much has it grown altogether?
   h How much does the cutting grow during the first three days?
   i How much does the cutting grow during the last three days?
   j On which day does the cutting grow exactly 3 cm?

2. The graph shows the depth of water in a harbour throughout the day. Answer the following questions using the graph.

   a What is the greatest depth of water in the harbour?
   b How deep is the water at 3 a.m.?
   c How deep is the water in the harbour at 8 p.m.?
   d At what time is high tide?

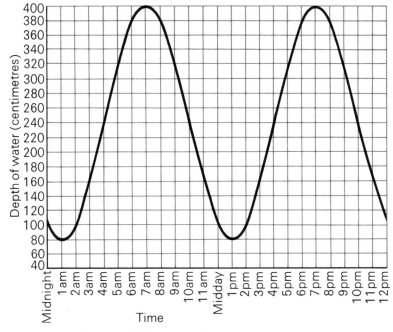

e At what time is low tide?

f When is the water in the harbour 240 cm deep?

g A yacht needs at least 320 cm of water in order to enter the harbour, between what times can it enter the harbour?

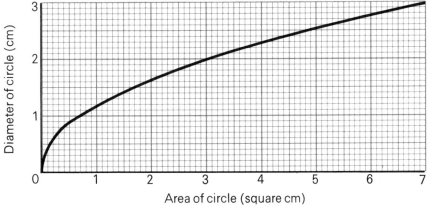

3 This graph gives the diameter of a circle if you are given its area. Find from the graph the diameters of circles with the following areas:

   a 4 sq cm   b 2 sq cm   c 1·5 sq cm
   d 5·3 sq cm   e 6·6 sq cm

Find the areas of the circles with the following diameters:

   f 1·5 cm   g 2·1 cm   h 1·8 cm   i 2·4 cm   j 2·9 cm

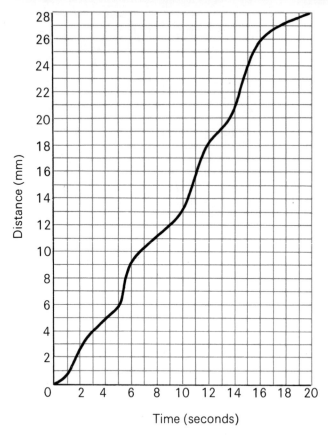

4 This graph shows the movement of a rotifer (a very small animal) on a microscope slide. Use the graph to answer these questions.
  a How far has the rotifer moved during the first 3 seconds?
  b How long does the rotifer take to move 21 mm from the start?
  c How far does the rotifer travel between 6 and 10 seconds?
  d At what time has the rotifer travelled 24 mm?
  e What distance does the rotifer travel during the 11th and 12th seconds?
  f When is the rotifer travelling slowest?
  g When is the rotifer travelling fastest?
  h How long does the rotifer take to increase the distance travelled from 13 mm to 18 mm?
  i How far does the rotifer travel in the first 10 seconds?
  j What is the average movement per second during the 20-second period?

Drawing graphs

Sometimes it is necessary to draw a graph if you are given a set of figures.

Example

The number of cars sold by a garage is shown in the table below. Draw a graph illustrating this, and use the graph to answer these questions:

1 When were the sales increasing fastest. From 1965 to 1969, or from 1969 to 1972?

2 In what year did the sales slow down?

3 Continue the line of the graph for the years 1973 and 1974, and use this line to predict **a** How many cars will be sold in 1973 **b** How many cars will be sold in 1974.

Year	1965	1966	1967	1968
Number of cars sold in each year	200	300	400	500

Year	1969	1970	1971	1972
Number of cars sold in each year	600	650	700	750

You will see that we need to allow space for ten years along the bottom of the graph, and about 10 spaces up the side, each space standing for 100 cars.

When the axes are drawn and marked with the years and the number of cars, find the line going up from 1965 and mark a dot where this line reaches 200, do the same for the other years, and then join the dots up. Notice carefully where the dots for 650 and 750 are marked.

Number of cars sold by a garage in years 1965 to 1972

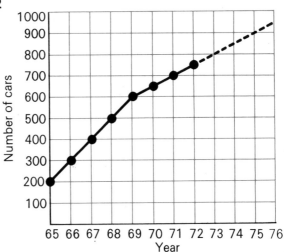

We can now use the graph to answer the questions.

1 You will see that the line of the graph is steepest between 1965 and 1969, so the sales were increasing faster between these years.

2 After 1969 the graph become less steep, and so the sales slowed down in 1970 when only 50 more cars were sold than in 1969.

3 The graph has been continued with a dotted line. If you read off the sales for 1973 and 1974 from the graph you get 800 cars sold in 1973 and 850 in 1974. It is of course impossible to say that these numbers of cars will be sold in these two years, but it will give the garage a rough idea of the number of cars it might expect to sell.

Exercise 20.2

1 The number of cars sold by a garage in certain years is given in the table below.

Year	1965	1967	1969	1971	1973
Number of cars sold in each year	300	400	500	600	700

On squared paper mark axes as in the example that has just been done for you, and mark in the points which show the number of cars sold in each year. Join the points up to make a graph, and use the graph to answer the following questions.

a How many cars were sold in 1966?

b How many cars were sold in 1968?

c How many cars were sold in 1972?

d In what year did the garage sell 550 cars?

e Are the answers you have just got completely accurate, or are they just rough estimates?

f In what year will the garage expect to sell 750 cars?

g How many cars can we estimate the garage will sell in 1975?

2 Repeat question 1 using the numbers in the following table. (The graph will not be a single straight line this time.)

Year	1965	1967	1969	1971	1973
Number of cars sold in each year	500	400	300	400	500

You will need to mark the years up to 1978.

Answer the following questions as well.

h Is the answer to **f** very reliable?

i Which year did they change the car salesman and why?

3 An experiment was carried out to see how long it took to melt different weights of steel in an electric furnace. The results are given in the table below.

Weight of metal (kg)	2	6	10	14
Time it takes to melt (minutes)	1	3	5	7

Mark weights up to 20 kg on the horizontal axis, and time up to 10 minutes on the vertical axis. Draw the graph of the results above and use it to answer the following questions.

a How long will it take to melt 4 kg?

b How long will it take to melt 8 kg?

c How long will it take to melt 20 kg?

d How long will it take to melt 11 kg?

e How long will it take to melt 17 kg?

f What weight can be melted in 6 minutes?

g What weight can be melted in 8 minutes?

h What weight can be melted in $6\frac{1}{2}$ minutes?

i What weight can be melted in $9\frac{1}{2}$ minutes?

4 A machine punches out metal discs. The weights of discs of different size are given in the table below.

Diameter (cm)	2	4	6	7	8	10
Weight (grammes)	0·4	1·6	3·6	4·9	6·4	10

Mark diameters up to 10 cm on the horizontal axis, and weights up to 10 g on the vertical axis. Draw a graph of the results and use it to answer the following questions. (The points will not be on a straight line, and you should carefully draw a curved line going through the points.) What are the weights of discs with the following diameters:

a 1 cm **b** 5 cm **c** 9 cm **d** 2·5 cm **e** 7·5 cm?

Give your answers to the nearest tenth of a gramme

What are the diameters of the discs with the following weights:

f 1 g **g** 5 g **h** 8 g **i** 7·5 g **j** 8·5 g?

Give your answers to the nearest tenth of a cm.

5 The time a car takes to go from London to Edinburgh at different speeds is shown in the table below.

Speed (km per hour)	60	80	100	120	140	160
Time taken (hours)	10·7	8	6·4	5·3	4·6	4

On the horizontal axis mark the speed from 60 km per hour to 160 km per hour, using one square for every 10 km per hour.

On the vertical axis mark the time taken from 4 hours to 11 hours, using one square for each hour.

Draw the graph of the results above and use it to answer the following questions. (The points will not be on a straight line, and you should carefully draw a curved line going through the points.)

How long will it take at the following speeds:

a 70 km per hour **b** 90 km per hour
c 110 km per hour **d** 130 km per hour
e 150 km per hour?

Give your answers to the nearest tenth of an hour.

What speed will be needed if the journey is to take the following times:

f 9 hours **g** 7 hours **h** 6 hours **i** 5 hours?

Give your answers to the nearest 1 km per hour.

In the graphs we have drawn so far the table of numbers used to draw the graph has been given. Sometimes the table has to be calculated before the graph can be drawn.

Example

2 metres of steel pipe weighs 3 kg. Draw a graph that will give the weights of up to 10 metres of steel pipe.

We know that   2 metres weigh 3 kg,
         so   4 metres weigh 6 kg,
              6 metres weigh 9 kg,
              8 metres weigh 12 kg,
             10 metres weigh 15 kg.

We can now put this in a table.

Length of pipe (metres)	2	4	6	8	10
Weight of pipe (kg)	3	6	9	12	15

Example

When the points are plotted on the graph you will see that they lie on a straight line which can be drawn with a ruler.

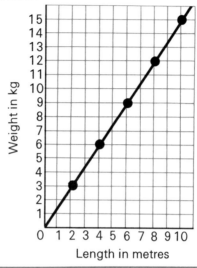

Exercise 20.3

1 3 m of steel pipe weighs 4 kg. Copy and complete this table.

Length of pipe (m)	3	6	9	12	15
Weight of pipe (kg)	4	8			

Use the table to draw a graph, and answer the following questions.

Find the weights of the following lengths of pipe:

a 2 m   b 5 m   c 8 m   d 11 m   e 14 m

Give your answers to the nearest tenth of a kg.

Find the lengths of pipe with the following weights:

f 3 kg   g 5 kg   h 9 kg   i 17 kg   j 19 kg

Give your answers correct to the nearest tenth of a metre.

2 A gardening shop sells fertilizer at 10p for 3 kg. Copy and complete this table.

Weight of fertilizer (kg)	3	6	9	12	15
Cost (pence)	10				

Use the table to draw a graph, and answer the following questions.

Find the cost of the following weights of fertilizer:

a 4 kg   b 7 kg   c 10 kg   d 13 kg   e 14 kg

Give your answer to the nearest 1p.

How much fertilizer can you buy for:

f 15p **g** 25p **h** 35p **i** 5p **j** 45p?

Give your answer to the nearest tenth of a kg.

3 The distance around a circle is given by the following formula:

Distance around = 3·1 × Diameter.

Copy and complete this table.

Diameter (cm)	2	4	6	8	10
Distance around (cm)	6·2				

Use the table to draw a graph, and answer the following questions.

What is the distance around a circle with the following diameters:

a 1 cm **b** 3 cm **c** 5 cm **d** 7 cm **e** 9 cm?

Give your answers to the nearest tenth of a cm.

Find the diameters of circles, the distance around which are:

f 5 cm **g** 8 cm **h** 15 cm **i** 19 cm **j** 27 cm

Give your answers to the nearest tenth of a cm.

4 Copy and complete this table which gives the areas of squares with sides up to 10 cm.

Length of side (cm)	0	1	2	3	4	5	6	7	8	9	10
Area (sq cm)	0	1	4	9							

Use the table to draw a graph. (The points will not be on a straight line, and you should carefully draw a curved line going through the points.)

Use the graph to answer the following questions.

What is the area of a square with a side of:

a 2·5 cm **b** 4·5 cm **c** 6·5 cm **d** 7·5 cm
e 8·5 cm?

Give your answer to the nearest square cm.

What is the length of the side of a square with an area of:

f 12 sq cm **g** 20 sq cm **h** 57 sq cm
i 75 sq cm **j** 85 sq cm?

5 The time it takes to go 48 km is given by the formula:

Time = 48 ÷ speed

Copy and complete this table.

Speed (km per hour)	2	3	4	6	8	12	16	24	48
Number of hours taken to go 48 km	24	16							

Use the table to draw a graph, and use the graph to answer the following questions.

How long will it take to go 48 km at the following speeds:

a 5·5 km per hour   b 7 km per hour
c 9 km per hour   d 14 km per hour
e 20 km per hour?

What speed do you need to go to travel 48 km in:

f 20 hours   g 15 hours   h 10 hours
i $2\frac{1}{2}$ hours   j $1\frac{1}{2}$ hours?

unit 21 Matrices

If a teacher gives tests in English, Mathematics, Science, and French to three boys he could write down the marks out of ten like this.

	John	Brian	Peter
English	5	7	6
Mathematics	4	5	3
Science	8	9	6
French	4	8	3

Another way of writing the marks down is like this:

$$\begin{bmatrix} 5 & 7 & 6 \\ 4 & 5 & 3 \\ 8 & 9 & 6 \\ 4 & 8 & 3 \end{bmatrix}$$

A set of numbers like this is called a **matrix**. This matrix is called a 4 by 3 matrix because it has 4 rows and 3 columns. This is called the size or order of the matrix.

The next week the teacher set four more tests. The matrix of the marks was

$$\begin{bmatrix} 6 & 6 & 7 \\ 3 & 5 & 4 \\ 9 & 8 & 5 \\ 5 & 7 & 4 \end{bmatrix}$$

What mark did Brian get on his second English test?
What mark did Peter get on his second Science test?
What mark did John get on his second French test?
How many marks did Brian get on all of his second tests?
What was the total of their marks on the second French test?
Which French test was the harder?

Exercise 21.1

1 Four boys took tests in English, Mathematics, and Science. The marks out of ten on the two sets of tests are shown below.

Test 1

	Peter	John	Brian	Victor
English	7	6	9	2
Maths	8	4	8	5
Science	9	5	10	3

Test 2	Peter	John	Brian	Victor
English	8	7	8	3
Maths	6	5	9	4
Science	8	6	9	4

a How many rows has each matrix?
b How many columns has each matrix?
c What is the size or order of each matrix?
d Each of the numbers in a matrix is called an **element**. How many elements are there in each matrix?
e What mark did Victor get on his first Maths test?
f What mark did Peter get on his second Science test?
g Who got the smallest English mark in the first test?
h Who got the largest mark for Science in the second test?
i What was the total of Brian's marks on the second test?
j What was the total of all the English marks on the first test?
k What was the total of John's Science marks in both tests?
l Who came top in the first test?
m Who came bottom in the second test?
n Who came top when the marks for both tests were added?

2 $\begin{bmatrix} 3 & 8 \\ 5 & 9 \\ 3 & 8 \\ 2 & 6 \end{bmatrix}$

a How many rows has this matrix?
b How many columns has this matrix?
c What is the size or order of this matrix?
d How many elements are there in this matrix?
e What number is in the first row and second column?
f What number is in the fourth row and first column?

3 Answer the same questions as in question **2** using this matrix.

$$\begin{bmatrix} 4 & 6 & 7 & 8 \\ 2 & 0 & 8 & 8 \\ 2 & 8 & 4 & 9 \end{bmatrix}$$

4 Answer the same questions as in question **2** using this matrix.

$$\begin{bmatrix} 5 & 6 & 7 & 9 \end{bmatrix}$$

Adding and taking away matrices

If the teacher wanted to add together the two sets of test marks on page 158 he could do it like this:

$$\begin{bmatrix} 5 & 7 & 6 \\ 4 & 5 & 3 \\ 8 & 9 & 6 \\ 4 & 8 & 3 \end{bmatrix} + \begin{bmatrix} 6 & 6 & 7 \\ 3 & 5 & 4 \\ 9 & 8 & 5 \\ 5 & 7 & 4 \end{bmatrix} = \begin{bmatrix} 11 & 13 & 13 \\ 7 & 10 & 7 \\ 17 & 17 & 11 \\ 9 & 15 & 7 \end{bmatrix}$$

This is called adding the two matrices. The new matrix contains the results of both the tests. For example it shows that Brian's French marks in the two tests came to 15 out of 20.

You can also take away one matrix from another like this:

$$\begin{bmatrix} 13 & 26 \\ 15 & 27 \end{bmatrix} - \begin{bmatrix} 10 & 20 \\ 5 & 25 \end{bmatrix} = \begin{bmatrix} 3 & 6 \\ 10 & 2 \end{bmatrix}$$

Matrices can only be added or subtracted if they are the same size. Examples like the following cannot be done:

$$\begin{bmatrix} 3 & 8 \\ 5 & 7 \\ 2 & 9 \end{bmatrix} + \begin{bmatrix} 6 & 4 \\ 3 & 8 \end{bmatrix}$$

Multiplying a matrix by a number

A matrix can also be multiplied by a number. If the teacher wanted to change the marks on the first test to marks out of 100 instead of marks out of ten he would have to make each mark ten times as big. He could do this by multiplying the matrix by 10 like this:

$$10 \times \begin{bmatrix} 5 & 7 & 6 \\ 4 & 5 & 3 \\ 8 & 9 & 6 \\ 4 & 8 & 3 \end{bmatrix} = \begin{bmatrix} 50 & 70 & 60 \\ 40 & 50 & 30 \\ 80 & 90 & 60 \\ 40 & 80 & 30 \end{bmatrix}$$

Exercise 21.2

1 Copy and complete these three examples.

a $\begin{bmatrix} 2 & 3 & 4 \\ 3 & 7 & 9 \\ 6 & 2 & 8 \end{bmatrix} + \begin{bmatrix} 5 & 6 & 8 \\ 2 & 9 & 3 \\ 8 & 4 & 0 \end{bmatrix} = \begin{bmatrix} 7 & 9 & \\ & & \\ & & \end{bmatrix}$

b $\begin{bmatrix} 16 & 8 \\ 9 & 7 \\ 12 & 8 \end{bmatrix} - \begin{bmatrix} 10 & 7 \\ 2 & 7 \\ 10 & 5 \end{bmatrix} = \begin{bmatrix} 6 & 1 \\ & \\ & \end{bmatrix}$

c $3 \times \begin{bmatrix} 4 & 6 & 7 \\ 3 & 5 & 1 \\ 4 & 0 & 6 \end{bmatrix} = \begin{bmatrix} 12 & 18 & \\ & & \\ & & \end{bmatrix}$

2 Add these matrices. If any of them cannot be done say so.

a $\begin{bmatrix} 2 & 4 \\ 3 & 6 \end{bmatrix} + \begin{bmatrix} 6 & 7 \\ 7 & 8 \end{bmatrix}$

b $\begin{bmatrix} 3 & 5 & 7 \\ 2 & 8 & 3 \end{bmatrix} + \begin{bmatrix} 2 & 4 & 9 \\ 7 & 7 & 8 \end{bmatrix}$

c $\begin{bmatrix} 5 \\ 4 \\ 5 \\ 4 \end{bmatrix} + \begin{bmatrix} 9 \\ 7 \\ 8 \\ 8 \end{bmatrix}$

d $\begin{bmatrix} 2 & 4 & 5 \\ 3 & 0 & 8 \end{bmatrix} + \begin{bmatrix} 5 & 8 \\ 3 & 8 \\ 3 & 9 \end{bmatrix}$

3 Subtract these matrices. If any of them cannot be done say so.

a $\begin{bmatrix} 9 & 6 & 8 \\ 3 & 4 & 9 \\ 2 & 2 & 7 \end{bmatrix} - \begin{bmatrix} 4 & 4 & 3 \\ 3 & 3 & 9 \\ 2 & 1 & 5 \end{bmatrix}$

b $\begin{bmatrix} 3 & 7 & 8 \end{bmatrix} - \begin{bmatrix} 4 \\ 3 \\ 9 \\ 7 \end{bmatrix}$

c $\begin{bmatrix} 45 & 27 \\ 26 & 34 \end{bmatrix} - \begin{bmatrix} 13 & 22 \\ 17 & 13 \end{bmatrix}$

d $\begin{bmatrix} 7 & 9 \\ 2 & 7 \\ 2 & 0 \end{bmatrix} - \begin{bmatrix} 10 & 8 \\ 6 & 9 \\ 3 & 5 \end{bmatrix}$

4 Multiply these matrices by the numbers shown.

a $2 \times \begin{bmatrix} 4 & 5 & 7 \\ 3 & 6 & 9 \\ 4 & 9 & 5 \end{bmatrix}$ b $3 \times \begin{bmatrix} 5 & 7 \\ 4 & 6 \\ 3 & 6 \end{bmatrix}$

c $9 \times \begin{bmatrix} 23 & 14 \\ 13 & 25 \\ 17 & 3 \end{bmatrix}$ d $\tfrac{1}{2} \times \begin{bmatrix} 34 & 25 & 16 \\ 12 & 0 & 17 \end{bmatrix}$

5 A baker delivers bread on Mondays, Wednesdays, and Fridays. The deliveries to three houses in a road on a Monday are given in the matrix below.

	Brown Bread	White Bread	Bread Rolls
Number 1	0	3	12
Number 4	1	1	0
Number 6	2	2	16

The deliveries on Wednesday and Friday are given in the matrices below.

$\begin{bmatrix} 0 & 2 & 10 \\ 2 & 1 & 0 \\ 2 & 1 & 20 \end{bmatrix} \quad \begin{bmatrix} 0 & 4 & 17 \\ 2 & 2 & 0 \\ 3 & 1 & 24 \end{bmatrix}$

a Write down the matrix which gives the deliveries for Monday and Wednesday.

b Write down the matrix which gives the deliveries for the whole week.

c If the deliveries for the following week were exactly the same, write down the matrix that gives the deliveries for the two weeks.

6 A man owns two garages. The amount of petrol (in litres) sold on various days is given in the matrices shown.

	Monday 1st Garage	Monday 2nd Garage	Tuesday		Tuesday and Wednesday	
3 Star	20	30	30	40	80	90
4 Star	35	40	60	50	85	120
5 Star	25	35	15	45	55	65

a Write down a matrix showing the amount of petrol sold on Monday and Tuesday.

b Write down a matrix showing the amount sold on Monday, Tuesday, and Wednesday.

c Write down a matrix showing the amount sold on Wednesday.

7 In question **6** the man finds that he usually sells twice as much petrol on Saturday as on Monday. On Sunday he sells half the petrol he sells on Monday.

a Write down a matrix showing the amount of petrol he expects to sell on Saturday.

b Write down a matrix showing the amount of petrol he expects to sell on Sunday.

Sometimes letters are used to stand for matrices so that they need not be written out in full.

Example

If $A = \begin{bmatrix} 3 & 4 & 5 \\ 2 & 6 & 7 \end{bmatrix}$ $B = \begin{bmatrix} 1 & 2 & 1 \\ 2 & 5 & 3 \end{bmatrix}$

$C = \begin{bmatrix} 10 & 14 & 12 \\ 13 & 15 & 23 \end{bmatrix}$ $D = \begin{bmatrix} 3 & 8 \\ 5 & 7 \end{bmatrix}$

Work out **1** $A+B$ **2** $B+C$ **3** $C-D$
         4 $C-B$ **5** $D+A$

1 $A+B = \begin{bmatrix} 4 & 6 & 6 \\ 4 & 11 & 10 \end{bmatrix}$

2 $B+C = \begin{bmatrix} 11 & 16 & 13 \\ 15 & 20 & 26 \end{bmatrix}$

3 Cannot be done because C and D are not the same size.

4 $C - B = \begin{bmatrix} 9 & 12 & 11 \\ 11 & 10 & 20 \end{bmatrix}$

5 Cannot be done because D and A are not the same size.

Example

Work out $4 \times A$ and $2D$.

$4 \times A = \begin{bmatrix} 12 & 16 & 20 \\ 8 & 24 & 28 \end{bmatrix}$ $\qquad 2D = 2 \times D = \begin{bmatrix} 6 & 16 \\ 10 & 14 \end{bmatrix}$

Exercise 21.3

$A = \begin{bmatrix} 2 & 4 & 8 \\ 3 & 7 & 6 \end{bmatrix} \qquad B = \begin{bmatrix} 1 & 2 & 5 \\ 1 & 5 & 3 \end{bmatrix}$

$C = \begin{bmatrix} 5 & 7 \\ 8 & 9 \\ 5 & 2 \end{bmatrix} \qquad D = \begin{bmatrix} 3 & 5 \\ 7 & 9 \\ 4 & 1 \end{bmatrix}$

Work out the following. If they cannot be done say so.

1 $A + B$ 2 $C + D$ 3 $B + D$
4 $A - B$ 5 $C - D$ 6 $3 \times A$
7 $2A$ 8 $3B$ 9 $2A + 3B$
10 $B - A$ 11 $D - C$ 12 $4C$
13 $2D$ 14 $4C - 2D$ 15 $\frac{1}{2}D$

unit 22 Shapes that are the Same

In mathematics we call shapes that are exactly the same **congruent**.

Underneath are some shapes that are **congruent**.

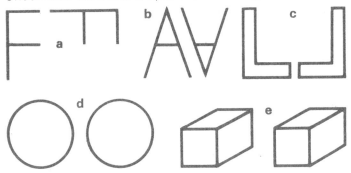

You will see that shapes that are congruent sometimes have to be turned around before they can fit over each other exactly as in **a** and **b**. They may even have to be turned over as in **c**. Solid shapes can of course be congruent like the boxes in **e**.

One way of seeing if flat shapes are congruent is by copying one using tracing paper, and seeing if you can make it fit exactly over the other shape.

Here are some shapes that are not congruent.

Exercise 22.1

Which of the following pairs of shapes are congruent? You may need tracing paper to do some of them.

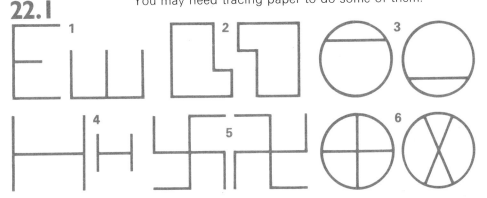

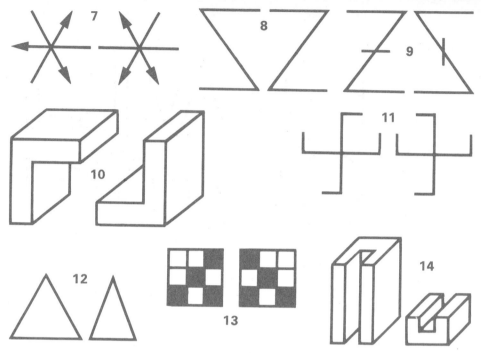

Enlargement and similar figures

Draw an arrow and mark the ends A and B. Mark a point by the side of the arrow and label it O. Your drawing should look like this. (Make AB a whole number of cm, say 1 cm, 2 cm, or 3 cm.)

Draw two lines through O and A and O and B. Your drawing should look like this.

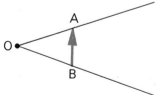

Mark a point on the line OA with the letter A_1 so that distance OA = distance AA_1.

Mark a point on the line OB with the letter B_1 so that distance OB = distance BB_1.

(The best way of making sure that the distances are equal is either to use a pair of compasses, or mark the distance OA on the edge of a piece of paper, and then use the marks to make sure that the distance AA_1 is the same.)

unit 22/page 167

Your drawing should look like this.

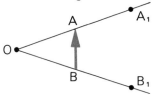

Now draw the arrow from A_1 to B_1. Your drawing should now look like this.

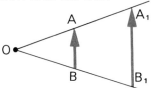

Measure the distance AB. Measure the distance A_1B_1. What do you notice about the two distances?

You should find that A_1B_1 is twice as long as AB.

What you have done is to draw an **enlargement** of AB.

O is called the **centre of enlargement**.

Since the new shape is twice the size of the first one we say that the enlargement has a **scale factor** of 2.

Draw a triangle and mark the corners A, B, C. Mark a point O outside the triangle. Draw three lines from O through A, B, and C.

Mark in points A_1, B_1, and C_1 so that OA_1 = three times OA, OB_1 = three times OB, and OC_1 = three times OC.

Draw the triangle $A_1B_1C_1$. Your drawing should look something like this.

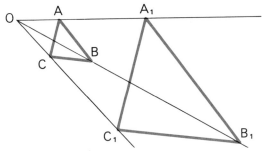

Measure the sides of the two triangles. What do you notice about their lengths? You should find that the sides of triangle $A_1B_1C_1$ are three times as long as the sides of triangle ABC.

This is because $OA_1 = 3 \times OA$, $OB_1 = 3 \times OB$, $OC_1 = 3 \times OC$.

We have enlarged the triangle with a scale factor of three.

Here is a four sided shape that has been enlarged with a scale factor of 2. In this case you will see that we have put the centre of enlargement inside.

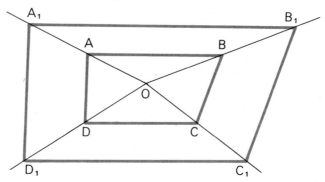

Exercise 22.2

Draw some shapes of your own and enlarge them with scale factors 2 or 3. (You can use scale factors of more than 3, but you will need large sheets of paper.)

Here are some shapes to give you some ideas.

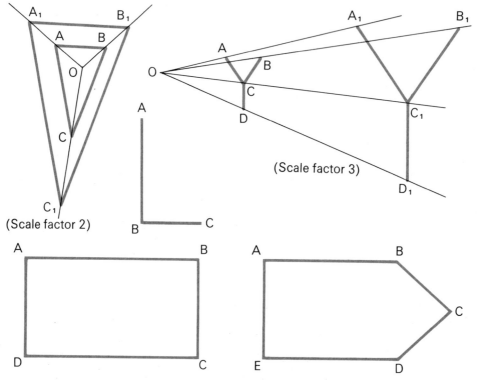

These two shapes are best enlarged with the centre of enlargement inside.

Enlargements using co-ordinates

Look at the rectangle ABCD. The co-ordinates of the four corners are A (2, 6)   B (4, 6)   C (4, 1)   D (2, 1).

We can write them in a matrix like this

$$\begin{bmatrix} 2 & 6 \\ 4 & 6 \\ 4 & 1 \\ 2 & 1 \end{bmatrix}$$

Let us multiply the matrix by 3

$$3 \times \begin{bmatrix} 2 & 6 \\ 4 & 6 \\ 4 & 1 \\ 2 & 1 \end{bmatrix} = \begin{bmatrix} 6 & 18 \\ 12 & 18 \\ 12 & 3 \\ 6 & 3 \end{bmatrix}$$

Let us now mark in the four new points A_1, B_1, C_1, D_1 where A_1 is (6, 18) and so on.

You will see that the new shape is still a rectangle, and if you count the number of squares along each side you will see that the new rectangle is 6 by 15 and the old rectangle is 2 by 5. You will see that the rectangle has been enlarged by a scale factor of 3. You will see also from the diagram that the centre of enlargement is at (0, 0).

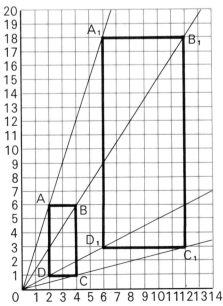

Exercise 22.3

For each of these examples you will need squared paper with numbers up to 20 on both axes.

1. **a** Mark in the points A (1, 4), B (1, 1), C (3, 1) join them to make the letter L.

   b Copy and complete this matrix multiplication.

   $$4 \times \begin{bmatrix} 1 & 4 \\ 1 & 1 \\ 3 & 1 \end{bmatrix} = \begin{bmatrix} & \\ & \\ & \end{bmatrix}$$

   c Mark in the three new points A_1, B_1, C_1. You will see that you have enlarged the shape with a scale factor of 4.

2. **a** Mark in the points A (3, 1), B (5, 4), C (3, 5), D (1, 4) and join them to make a kite shape.

   b Write down the matrix of the four points and multiply it by 3.

   c Mark in the four new points A_1, B_1, C_1, D_1, and join them to make a kite shape.

   d Check that the lines from O to A, B, C, D go through A_1, B_1, C_1, D_1.
   You will see that you have enlarged the shape with a scale factor of 3.

3. **a** Mark in the points A (3, 3), B (4, 5), C (5, 7), D (6, 5), E (7, 3) and join them to make a letter A.

   b Write down the matrix of the five points and multiply it by 2.

   c Mark in the new points A_1, B_1, C_1, D_1, E_1, and join them to make a letter A.
   You will see that you have enlarged the shape with a scale factor of 2.

4. **a** Mark in the points A (6, 6), B (12, 6), C (12, 12), D (6, 12), and join them to make a square.

   b Copy and complete this matrix multiplication.

   $$\tfrac{1}{2} \times \begin{bmatrix} 6 & 6 \\ 12 & 6 \\ 12 & 12 \\ 6 & 12 \end{bmatrix} = \begin{bmatrix} 3 & 3 \\ & \\ & \\ & \end{bmatrix}$$

   c Mark in the new points A_1, B_1, C_1, D_1, and join them to make a square. You will see that we now get a square that is smaller than our first square. If you measure the length of the sides of the squares you will see that the second square is $\frac{1}{2}$ the size of the first square. We say that we have enlarged the square with a scale factor of $\frac{1}{2}$.

5 Do number **4** again using $\frac{1}{3}$ instead of $\frac{1}{2}$ in the matrix multiplication. You will find that you have enlarged the square with a scale factor of $\frac{1}{3}$. (This means that the new square is $\frac{1}{3}$ the size of the first one.)

6 Mark in some shapes of your own and enlarge them by multiplying the matrix of their points by a scale factor. (Use scale factors less than 4 otherwise you will need a very large sheet of paper.)

In the next examples you will need squared paper with numbers from -10 to 10 on each axis.

7 a Mark in the points A (1, 4), B (2, 5), C (3, 4), D (2, 1), and join them to make an arrow.

b Copy and complete this matrix multiplication.

$$^{-}1 \times \begin{bmatrix} 1 & 4 \\ 2 & 5 \\ 3 & 4 \\ 2 & 1 \end{bmatrix} = \begin{bmatrix} ^{-}1 & ^{-}4 \\ & \\ & \\ & \end{bmatrix}$$

c Mark in the new points A_1, B_1, C_1, D_1, and join them to make an arrow. (A_1 is the point $(^{-}1, ^{-}4)$ and so on.)

d Draw 4 lines from A to A_1, from B to B_1, from C to C_1, from D to D_1. The lines should all pass through (0, 0). What you have done is to enlarge the shape with centre of enlargement (0, 0) and a scale factor of $^{-}1$. The minus sign in the scale factor turns the shape upside down, and the 1 means that the size is still the same.

8 Do the same as in question **7** with a scale factor of $^{-}2$. You should find that the arrow is turned upside down, and is twice as big.

page 172/unit 22

a

b

Exercise 22.4

Some shapes, like those shown above, are difficult to enlarge using centres of enlargement, or matrices. They are best enlarged by copying each part of the shape into bigger squares.

In the example shown above **a** has been enlarged into **b** with a scale factor of 2. The squares in **b** have been made twice as big as the squares in **a,** and each part of **a** has been drawn twice as big in the squares of **b**

1. On squared paper make a copy of **a**. It need not be the same size.
2. Now draw some squares twice as big as the squares you have used in **1**, and draw the shape twice as large.
3. Draw a shape of your own on squared paper.
4. Using squares twice as big draw an enlargement with scale factor 2.
5. Using squares three times as big draw an enlargement with scale factor 3.
6. Using squares half as big as those you have used in **3**, draw an enlargement with scale factor $\frac{1}{2}$.

Similar shapes

When a shape is enlarged, the two shapes you get are called **similar**. Two shapes are similar if you can enlarge one of them until it fits exactly over the other one.

Sometimes if two shapes are similar it is useful to be able to work out the measurements of one of them if you are given the measurements of the other one.

Example

The scale factor of this enlargement is 3. What are the measurements of the large shape?

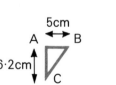
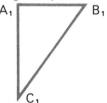

Since the scale factor is three, every length in the new shape is three times the length in the first shape. So we can say:

$A_1 B_1 = 3 \times 5 = 15$ cm

$A_1 C_1 = 3 \times 6 \cdot 2 = 18 \cdot 6$ cm

Sometimes we have to work out the scale factor first.

Example

Find the scale factor, and then the distance B_1C_1.

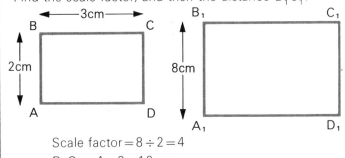

Scale factor $= 8 \div 2 = 4$
$B_1C_1 = 4 \times 3 = 12$ cm

Example

An exact copy of this letter L has to be made, enlarged so that the base is 16 cm long. What are the other measurements of the enlarged letter?

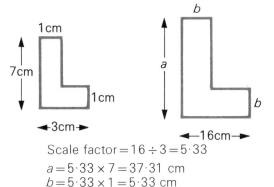

Scale factor $= 16 \div 3 = 5.33$
$a = 5.33 \times 7 = 37.31$ cm
$b = 5.33 \times 1 = 5.33$ cm

Instead of saying that there is a scale factor for example of 10, we can say instead that there is a scale of 1 to 10.

Exercise 22.5

A piece of metal has to be cut similar to the one shown (7 cm by 12 cm).

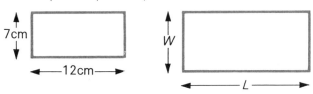

1 Find the lengths W and L if the scale factor is 2.
2 Find the lengths W and L if the scale factor is 3.
3 Find the lengths W and L if the scale factor is 7.
4 Find the lengths W and L if the scale factor is 9.

Find the lengths W and L if the scale factor is:
5 12 **6** 13 **7** 2.5 **8** 3.5 **9** 3.75 **10** 13.5

A manufacturer wishes to make boxes similar to the one shown below. Find the lengths L, W, and H if the scale factor is:

11 4   12 7   13 ½   14 0·3   15 0·7

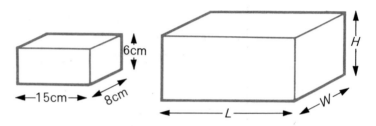

In each of the following examples find the scale factor, and then find the lengths marked by the letters.

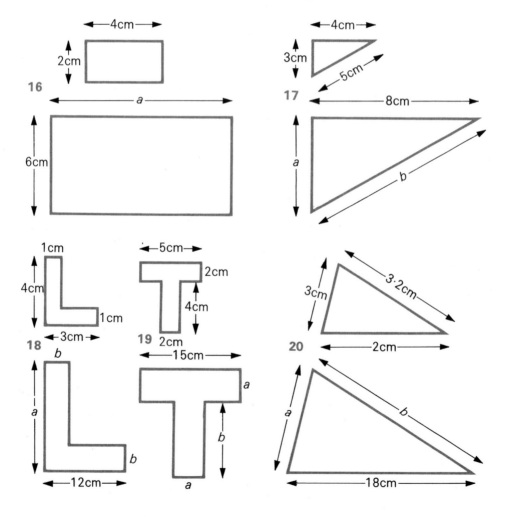

In each of the following examples find the scale factor, and then find the lengths a, b, and c.

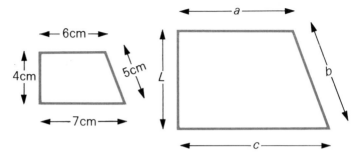

21 If L = 10 cm   22 If L = 13 cm   23 If L = 6 cm
24 If L = 2 cm   25 If L = 1 cm

26 A toy bus is 7·8 cm long. It is made on a scale of 1 to 100.
   a What is the length of the full size bus in cm?
   b What is the length of the full size bus in m?

27 A car is 3 m long, 2 m wide, and 1·5 m high. A model is to be made on a scale of 1 to 10.
   a What are the measurements of the car in cm?
   b What are the measurements of the model in cm?

28 A plan of a factory is made on a scale of 1 to 2000.
   a What is the size of the factory if the plan is 30 cm by 40 cm? Give your answer in m.
   b What is the size of the plan if the factory is 200 m by 500 m? Give your answer in cm.

29

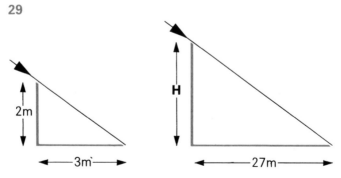

A stick 2 m high held in a vertical position has a shadow 3 m long. At the same time a building H m high had a shadow 27 m long. Find the height of the building. (Find the scale factor first.)

30 Find the height of the building in question 29 if the length of its shadow was 25 m.

answer section/page 177

unit 1

Exercise 1.1

	a	b	c	d	e	f	g	h	i
1	7	8	12	16	28	54	88	189	242
2	10	11	15	19	31	57	91	192	245
3	11	12	16	20	32	58	92	193	246
4	19	20	24	28	40	66	100	201	254
5	27	28	32	36	48	74	108	209	262
6	58	59	63	67	79	105	139	240	293
7	99	100	104	108	120	146	180	281	334
8	206	207	211	215	227	253	287	388	441
9	734	735	739	743	755	781	815	916	969
10	983	984	988	992	1004	1030	1064	1165	1218

11 910   **12** 1116   **13** 1833   **14** 1556   **15** 12 281
16 310 059

	a	b	c	d	e	f	g	h	i
1	3	2	2	6	18	44	78	179	232
2	6	5	1	3	15	41	75	176	229
3	7	6	2	2	14	40	74	175	228
4	15	14	10	6	6	32	66	167	220
5	23	22	18	14	2	24	58	159	212
6	54	53	49	45	33	7	27	128	181
7	95	94	90	86	74	48	14	87	140
8	202	201	197	193	181	155	121	20	33
9	730	729	725	721	709	683	649	548	495
10	979	978	974	970	958	932	898	797	744

17 5123   **18** 2189   **19** 3129   **20** 1049   **21** 21 813

Exercise 1.2

Many pupils have difficulty in translating a verbal or written problem into the arithmetic techniques used to solve the problem. With this exercise the pupils should first decide whether they need to add or subtract, and then do the calculation. A class discussion on how to do the first few problems will be found helpful.

1 25p   **2** 21p   **3** 87p   **4** 18   **5** 19 g   **6** 84 tonnes
7 49p   **8** 57p   **9** 58p   **10** £1   **11** 278   **12** 229
13 192   **14** 579   **15** 399   **16** 755   **17** £596
18 £4172   **19** £1350   **20** £671

page 178/answer section

Exercise 1.3

This contains nearly 200 examples on the multiplication and division of whole numbers. Numbers 11 and 12 give a single figure answer with no remainder. Numbers 13 and 14 give a single figure answer with a remainder. Numbers 15 and 16 give a two figure answer with no remainder. The rest of the examples give answers with two or more figures in the answer with or without a remainder.

	a	b	c	d	e	f	g	h	i
1	26	48	134	196	90	172	150	246	1092
2	65	120	335	490	225	430	375	615	2370
3	91	168	469	686	315	602	525	861	3822
4	104	192	536	784	360	688	600	984	4368
5	78	144	402	588	270	516	450	738	3276
6	377	696	1943	2842	1305	2494	2175	3567	15834
7	169	312	871	1274	585	1118	975	1599	7098
8	533	984	2747	4018	1845	3526	3075	5043	22386
9	728	1344	3752	5488	2520	4816	4200	6888	30576
10	10283	18984	52997	77518	35595	68026	59325	97293	431886

	a	b	c	d	e	f	g	h	i
11	8	8	8	5	7	7	8	8	5
12	7	4	3	10	5	9	7	6	9
13	8 r 2	8 r 1	8 r 3	5 r 2	7 r 7	7 r 3	8 r 3	8 r 5	5 r 2
14	7 r 2	4 r 4	3 r 3	10 r 3	5 r 2	9 r 3	7 r 1	6 r 3	9 r 4
15	17	23	13	25	13	19	17	16	13
16	13	14	18	25	24	32	29	27	27
17	24 r 1	13 r 2	17 r 1	13 r 4	13 r 5	18 r 3	24 r 5	32 r 2	28 r 6
18	18	25 r 1	12 r 1	34 r 6	26 r 5	14 r 8	29 r 16	26 r 6	20 r 7
19	20 r 33	15 r 29	21 r 38	13 r 34	11 r 36	36 r 1	12 r 47	10 r 27	5 r 3

20 18 r 144
21 12 r 391
22 102 r 545
23 66 r 620
24 152 r 3615

Exercise 1.4

Many pupils have difficulty in translating a verbal or written problem into the arithmetic techniques necessary to solve the problem. With this exercise the pupils should first decide whether they need to multiply or divide, and then do the calculation. A class discussion on how to do the first few problems will be found helpful.

1 161p **2** 28p **3** 225 **4** 657 kg **5** 32 **6** 612
7 38p **8** £8·37 **9** 31 **10** 15 tonnes **11** 59
12 58 **13** £10·14 **14** £14 **15** 78p **16** £459
17 45 **18** £4845 **19** £64 **20** £2071965

answer section/page 179

Exercise 1.5

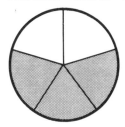

This contains 80 examples on changing from fractions to decimals and vice-versa. Before doing this work it may be necessary to ensure that the pupils understand the meaning of fractional notation, e.g. If 6 out of a class of 11 are girls, then $\frac{6}{11}$ of the class are girls. Or, $\frac{3}{5}$ of this circle (or cake) has been shaded (or cut).

	a	b	c	d	e
1	2·3	4·17	3·1	5·27	9·2
2	12·8	0·27	13·7	0·237	14·462
3	17·9	0·06	20·1	0·09	26·06
4	4·009	9·07	0·038	127·72	0·186
5	$3\frac{2}{10}$	$5\frac{37}{100}$	$4\frac{8}{10}$	$9\frac{64}{100}$	$9\frac{4}{10}$
6	$18\frac{2}{10}$	$\frac{75}{100}$	$19\frac{8}{10}$	$2\frac{87}{1000}$	$85\frac{543}{1000}$
7	$19\frac{7}{10}$	$\frac{3}{100}$	$43\frac{7}{10}$	$\frac{3}{10}$	$53\frac{6}{100}$
8	$9\frac{4}{1000}$	$9\frac{4}{100}$	$\frac{53}{1000}$	$149\frac{83}{100}$	$\frac{764}{1000}$

	f	g	h	i	j
1	7·05	6·92	10·06	8·43	11·07
2	0·6	15·3	0·471	0·5	16·4
3	0·2	9·3	0·039	38·42	41·064
4	8·043	0·271	7·027	11·006	63·042
5	$9\frac{5}{100}$	$9\frac{64}{100}$	$67\frac{5}{100}$		
6	$\frac{6}{10}$	$2\frac{5}{10}$	$\frac{635}{1000}$		
7	$\frac{9}{10}$	$6\frac{9}{10}$	$\frac{56}{1000}$		
8	$3\frac{95}{100}$	$\frac{345}{1000}$	$6\frac{23}{1000}$		

Exercise 1.6

	a	b	c	d	e	f	g
1	2·9	4·3	10·6	18·1	2·2	1·05	1·65
2	4·8	6·2	12·5	20	4·1	3·95	3·55
3	4·3	5·7	12	19·5	3·6	3·45	3·05
4	15	16·4	22·7	30·20	14·3	14·15	13·75
5	2	3·4	9·7	17·2	1·3	1·15	0·75
6	1·68	3·08	9·38	16·88	0·98	0·83	0·43
7	1·32	2·72	9·02	16·52	0·62	0·47	0·07
8	2·135	3·535	9·835	17·335	1·435	1·285	0·885
9	15·1	16·5	22·8	30·3	14·4	14·25	13·85
10	24·67	26·07	32·37	39·87	23·97	23·82	23·42

11 42·42 **12** 42·657 **13** 436·472 **14** 80·963 **15** 97·7078
16 113·586

page 180/answer section

	a	b	c	d	e	f	g
1	0·3	1·1	7·4	14·9	1	1·15	1·55
2	2·2	0·8	5·5	13	2·9	3·05	3·45
3	1·7	0·3	6	13·5	2·4	2·55	2·95
4	12·4	11	4·7	2·8	13·1	13·25	13·65
5	0·6	2	8·3	15·8	0·1	0·25	0·65
6	0·92	2·32	8·62	16·12	0·22	0·07	0·33
7	1·28	2·68	8·98	16·48	0·58	0·43	0·03
8	0·465	1·865	8·165	15·665	0·235	0·385	0·785
9	12·5	11·1	4·8	2·7	13·2	13·35	13·75
10	22·07	20·67	14·37	6·87	22·77	22·92	23·32

Exercise 1.7

Many pupils have difficulty in translating a verbal or written problem into the arithmetic techniques necessary to solve the problem. With this exercise the pupils should first decide whether they need to add or subtract, and then do the calculation. A class discussion on how to do the first few problems will be found helpful.

1 £3·82 **2** £0·59 **3** £1·94 **4** 85p **5** 4·70 m
6 10·1 m **7** 9·35 m **8** 14·6 tonnes **9** 1·89 kg
10 12·21 tonnes **11** 3·01 kg **12** 8·9 litres **13** £1·31
14 4·9 m **15** £0·53 **16a** £2·07 **b** £2·93
17 £2·67 **18** 6·5 m **19** 0·8 kg **20** 21·35 tonnes

unit 2

A suitable condensed treatment of the work in this Unit could be Exercise 1 together with the work on binary representation from Exercises 2, 3 and 4.

Exercise 2.1

This contains simple examples on changing numbers to and from bases other than ten. Most pupils should be able to do these without learning complex rules. Some pupils need go no further than this exercise, the purpose of which is to show that representation of numbers in other bases is possible.

	a	b	c	d	e	f
1	23_4	21_5	15_6	14_7	13_8	12_9
2	21_4	14_5	13_6	12_7	11_8	10_9
3	33_4	30_5	23_6	21_7	17_8	16_9
4	11_4	10_5	5_6	5_7	5_8	5_9
5	123_4	102_5	43_6	36_7	33_8	30_9
6	7	17	10	11	8	6
7	9	8	10	14	13	9
8	13	26	30	16	26	15
9	17	45	35	21	31	62
10	71	31	47	22	18	64

answer section/page 181

Exercise 2.2

1	422	**2**	566	**3**	237	**4**	236	**5**	366
6	2027	**7**	1225	**8**	2199	**9**	876	**10**	1891

	a	b	c	d	e	f	g
11	609	1243	106	230	249	158	58
12	49	70	16	3820	1451	501	1468
13	4360	696	745	2187	3932	182	3564
14	11	12	10	15	8	21	7

Exercise 2.3

	a	b	c	d	e
1	11231_4	23322_4	2022_4	23111_4	123320_4
2	604_7	1506_7	554_7	555_7	34326_7
3	2124_5	12103_5	3021_5	2134_5	120030_5
4	1553_6	3542_6	1454_6	3213_6	54122_6
5	435_8	1033_8	511_8	1343_8	5635_8
6	10111_2	101111_2	10000_2	101001_2	11110011_2

Exercise 2.4

This contains a more detailed treatment of binary numbers together with some information on the use of the binary system in computers.

1 B 19, C 15, D 20, E 31, F 35, G 39, H 60, I 42, J 64, K 80, L 74, M 85, N 91, O 95, P 127

2
a ○○○○●●○●
b ○○○●○○○●
c ○○○●○●●○
d ○○○●●○●●
e ○○○●●●●○
f ○○●○○○○●
g ○○●○○○●○
h ○○●○●○●●
i ○○●○●●●●
j ○○●●○●○○

3
a ○○●●●●○●
b ○●○○○○○●
c ○●○○○○●●
d ○●○○●○●○
e ○●○●○●●●
f ○●●○○●●●
g ○●●●○●○●
h ○●●●●○○○
i ○●●●●●○●
j ●○○○●●○○

unit 3

Exercise 3.1

1 {The days of the week}, 7.
2 {The last five letters of the alphabet}, 5.
3 {The even numbers from 2 to 10}, 5.
4 {The second five letters of the alphabet}, 5.

5 {The odd numbers from 1 to 9}, 5.
6 {The first five months of the year}, 5.
7 {The whole numbers from 95 to 100}, 6.
8 {The last four months of the year}, 4.
9 {v, w, x, y, z}.
10 {16, 17, 18, 19}.
11 {20, 21, 22, 23, 24, 25}.
12 {January, June, July}.
13 {3, 9, 15, 21, 27}.
14 {1970, 1971, 1972, 1973, 1974, 1975, 1976, 1977, 1978, 1979, 1980}.
15 {1, 2, 3, 4, 5, 6, 7, 8, 9}.

Exercise 3.2

1 {a, b, c, d, e}.
2 {Peter, Paul, Mary, John, George, Ringo}.
3 {1, 2, 3, 4, 5, 6, 7, 8, 10}.
4 {1, 2, 3, 5, 7}.
5 {2, 3, 4, 6, 8, 9, 10}.
6 {b, c, d}.
7 {Paul}.
8 {2}.
9 {3}.
10 {3, 5, 7}.
11 { }.
12 {1, 2, 3, 5, 6, 7, 9}.
13 { }.
14 {1, 2, 3, 5, 7}.
15 {2, 6}.
16 {2, 3, 6, 9}.

Exercise 3.3

1 a Mary, Peter; **b** Alan, Jane, Susan; **c** {Phillip}.
2 a Brown, Rogers, Smith, Singh; **b** Brown, Rogers.
3 a Mary, Jean; **b** John; **c** Joan, Gary.
4 a Roger, Susan; **b** Mary, Anne.
5 {4, 6, 8}.
6 a {January, July}; **b** January, July.
7 a l, m, i, a; **b** o, y, p, d.
8 a 6, 12, 18, 24, 30; **b** {30}.
9 a {11, 77}; **b** {1, 11, 21, 31, 41, 51, 61, 71, 77, 81, 91}.

answer section/page 183

Exercise 3.4

1 a 3; b 9; c 10; d 19.
2 a 25; b 21; c 12; d 13.
3 a 28
   b 3
   c 33

4 a 7
   b 8
   c 5

5 a 41; b 30; c 43; d 14; e 18; f 4.
6 a 19; b 23; c Humorous books and romances; d 12; e 5.
7 a 6; b 7; c 6; d 3.
8 a 9; b 6; c 58.
9 a 3; b 6; c 1.

unit 4

A suitable short treatment of this Unit would be Exercises 4.1 and 4.3, although Exercise 4.3 can be done by itself.

Exercise 4.1

	a	b		a	b
1	10, 12	16, 32	9	$4\frac{1}{2}$, 4	4, 8
2	39, 37	64, 128	10	7·5, 7·6	$\frac{1}{2}$, 1
3	26, 29	80, 160	11	17·6, 35·2	45·9, 45·5
4	21, 26	162, 486	12	31, 37	$\frac{1}{3}$, 1
5	33, 37	96, 192	13	113·6, 227·2	1, 5
6	26, 32	9, 3	14	16, 8	0·01, 0·001
7	13, 9	405, 1215	15	768, 3072	15, 5
8	37, 45	12, 4	16	2·9, 2·1	4, $4\frac{2}{3}$

Exercise 4.2

	1	2	3	4	5	6
a	16, 22	18, 24	31, 43	52, 40	46, 64	19, 8
b	37, 47	33, 42	12, 6	51, 67	38, 51	77, 107

	7	8	9	10
a	42, 59	66, 70	151·1, 211·1	9, 11
b	18, 15	101, 146	9, 12	10·8, 13·0

Exercise 4.3

This contains examples on the sums of even numbers, odd numbers, and whole numbers.

1. **e** 5×5; **f** 6×6; **g** 49=7×7; **h** 64; **i** 81; **j** 4; **k** 9; **l** 4×4=16; **m** 8×8=64; **n** 10×10=100; **o** 15×15=225
2. **e** 5×6; **f** 6×7; **g** 56=7×8; **h** 72; **i** 90; **j** 6; **k** 12; **l** 4×5=20; **m** 8×9=72; **n** 10×11=110; **o** 15×16=240
3. **e** 5×6; **f** $\frac{1}{2}$×42=$\frac{1}{2}$×6×7; **g** 28=$\frac{1}{2}$×56=$\frac{1}{2}$×7×8; **h** 36; **i** 45; **j** 3; **k** 6; **l** 10; **m** $\frac{1}{2}$×8×9=$\frac{1}{2}$×72=36 **n** $\frac{1}{2}$×10×11=$\frac{1}{2}$×110=55; **o** $\frac{1}{2}$×15×16=$\frac{1}{2}$×240=120

unit 5

Exercise 5.1

1. A=27°
2. B=43°
3. C=58°
4. D=71°
5. E=42°
6. F=133°
7. G=41°
8. H=13°
9. I=40°
10. J=323°
11. K=80°
12. L=197°
13. M=37°
14. N=57°
15. P=169°
16. Q=21°
17. R=19°
18. S=94°

Exercise 5.2

1. C, D, G=23°; A, B, E, F, H, I=157°.
2. K, L, M, P, N=130°; J=50°.
3. Q, R, U, V=112°; S, W, T=68°; X, Y, Z=51°.
4. A, B=112°; C=68°.
5. E=109°; D, F=71°.
6. H=73°; G, I=107°.
7. K, N=69°; J, L, M=111°.

Exercise 5.3

This exercise should lead pupils to discover that the angles of a triangle add up to 180°. It will also give them practice in measuring angles accurately.

	1	2	3
A	61°	64°	64°
B	57°	48°	42°
C	62°	68°	74°

4. A=102°
5. B=85°
6. C=95°
7. D=78°
8. E=75°
9. F=157°
10. G=60°
11. H=25°
12. I=15°
13. J=10°

To measure an angle when the arms of the angle do not reach to the graduations of the protractor, position the protractor in the usual way, and then align a ruler or the edge of a piece of paper along the arm to reach the graduations.

answer section/page 185

Exercise 5.4

This exercise should lead pupils to discover that the angles of a quadrilateral add up to 360°. It will also give them practice in measuring angles accurately.

	A	B	C	D
1	78°	109°	78°	95°
2	98°	104°	80°	78°

3 A = 89°
4 B = 107°
5 C = 39°
6 D = 57°

7 E = 78°
8 F = 64°
9 G = 75°
10 H = 63°

Exercise 5.5

This exercise overcomes the difficulty which arises when a class is given a number of angles to draw accurately as it can lead to problems of marking. With these examples the teacher or pupil can easily see if the angles are the correct size by checking the length of the line. Some pupils may need assistance with the diagrams for numbers 16 to 20.

1 2·0 cm	6 1·2 cm	11 8·0 cm	16 9·9 cm
2 2·8 cm	7 4·5 cm	12 8·8 cm	17 8·5 cm
3 5·5 cm	8 4·2 cm	13 9·8 cm	18 6·5 cm
4 6·7 cm	9 3·3 cm	14 9·9 cm	19 4·6 cm
5 5·2 cm	10 7·3 cm	15 9·9 cm	20 1·3 cm

unit 6

Care should be taken here that the pupil appreciates the difference between the two uses of the ' − ' sign. As an operation as in 6–9, and as an indication that a number is a negative number as in ⁻7. It could be pointed out here that positive numbers can be written not only as for example 15, but also as ⁺15. If an electronic calculating machine is available it can be used to demonstrate some of the problems, the minus sign when displayed indicating a loss, overdraft, etc.

Exercise 6.1

	a	b	c	d	e	f	g
1	⁻4	⁻3	⁻2	⁻2	⁻5	⁻2	4
2	⁻1	0	⁻7	⁻1	1	⁻2	⁻4
3	⁻1	5	⁻3	4	⁻5	⁻8	⁻4
4	⁻7	⁻4	⁻2	⁻5	⁻3	⁻6	⁻7
5	⁻2	⁻3	⁻1	⁻4			
6	⁻6	2	⁻7	⁻9			
7	1	⁻9	⁻2	⁻7			
8	3	0	⁻6	0			

Exercise 6.2

	a	b	c
1	⁻4	⁻10	4
2	⁻1	⁻9	1
3	⁻2	0	4
4	⁻6	⁻1	2

5 Overdraft of £6 **6** Overdraft of £8 **7** £3
8 Overdraft of £4 **9** £5 overdraft **10** £7 overdraft
11 £9 overdraft **12** £7 **13** £0 **14** £7 overdraft
15 £2 loss **16** £6 loss **17** £4 profit **18** £4 loss
19 £12 loss **20** £8 profit **21** £1·10 loss **22** £16·38

unit 7

Exercise 7.1

Numbers 11 and 12 come out exactly. Most of the remainder do not. This is a suitable point to introduce the concept of correcting up, although with some pupils it might be desirable at this stage to get them to divide until they have say 2 or 3 figures, and then introduce correcting up at a later stage.

	a	b	c	d	e	f
1	4·8	8·1	2·7	0·9	7·05	14·13
2	9·6	16·2	5·4	1·8	14·1	28·26
3	14·4	24·3	8·1	2·7	21·15	42·39
4	8	13·5	4·5	1·5	11·75	23·55
5	17·6	29·7	9·9	3·3	25·85	51·81
6	36·8	62·1	20·7	6·9	54·05	108·33
7	99·2	167·4	55·8	18·6	145·7	292·02
8	131·2	221·4	73·8	24·6	192·7	386·22
9	126·4	213·3	71·1	23·7	185·65	372·09
10	388·8	656·1	218·7	72·9	571·05	1144·53

	g
1	71·4
2	142·8
3	214·2
4	119
5	261·8
6	547·4
7	1475·6
8	1951·6
9	1880·2
10	5783·4

answer section/page 187

	a	b	c	d	e	f	g
11	0·7	1·3	0·8	1·8	2·3	0·8	1·1
12	6·2	0·43	0·13	1·69	14·2	2·6	0·54
13	11·53	6·4	1·49	1·39	3·695	3·206	9·422
14	0·195	0·05571	0·1732	1·942	2·088	5·966	4·222
15	2·88	0·5916	1·208	3·014	0·7417	0·02294	0·05194
16	0·5118	0·1617	10·38	3·792	27·17	0·2666	6·515
17	0·4807	6·863	2·094	1·096	0·03187	0·04016	0·7025
18	0·4313	0·1906	3·044	0·3135	0·01791	0·0003269	0·3571

Exercise 7.2

Many pupils have difficulty in translating a verbal or written problem into the arithmetic techniques necessary to solve the problem. With this exercise the pupils should first decide whether they need to multiply or divide, and then do the problems. A class discussion on how to do the first few problems will be found helpful.

1 £6·20
2 £0·64
3 10·92t
4 £2·88
5 £1·36
6 £0·57
7 £6·15
8 £8·04
9 5·6t
10 0·93 m
11 £6·08
12 £25·45½
13 £33·75
14 £1·79
15 £90·61
16 £0·95
17 £0·24
18 687·3 m

Exercise 7.3

Numbers 11 to 17 come out exactly. The remainder do not.

	a	b	c	d	e	f	g
1	0·3	0·35	0·20	0·55	1·15	1·75	0·365
2	0·42	0·49	0·28	0·77	1·61	2·45	0·511
3	0·48	0·56	0·32	0·88	1·84	2·8	0·584
4	0·78	0·91	0·52	1·43	2·99	4·55	0·949
5	1·44	1·68	0·96	2·64	5·52	8·4	1·752
6	0·276	0·322	0·184	0·506	1·058	1·61	0·3358
7	0·804	0·938	0·536	1·474	3·082	4·69	0·9782
8	3·786	4·417	2·524	6·941	14·513	22·085	4·6063
9	0·2076	0·2422	0·1384	0·3806	0·7958	1·211	0·25258
10	14·04	16·38	9·36	25·74	53·82	81·9	17·082

	g
1	0·365
2	0·511
3	0·584
4	0·949
5	1·752
6	0·3358
7	0·9782
8	4·6063
9	0·25258
10	17·082

	a	b	c	d
11	12	8	6	40
12	12	9	6	120
13	3·6	36	1·8	48
14	240	3600	18000	12000
15	1·2	0·6	6	12
16	0·12	1·1	0·11	1·2
17	13	130	15	150
18	21·76	23·12	1608	18·40
19	11·16	57·04	1069	12·58
20	2629	27937	732·7	85·14

Exercise 7.4

See the note on Exercise 7.2.

1 £2·70
2 13
3 38
4 £0·16
5 20
6 14
7 5·44
8 20
9 £0·17½
10 20
11 £1·87
12 15
13 18
14 £4·88½
15 £0·42
16 16·25
17 £20·54
18 36 (36·18)
19 £6·94½
20 29 (29·17)

Exercise 7.5

	a	b	c	d	e	f
1	27	270	2700	0·27	0·027	0·0027
2	34·5	345	3450	0·345	0·0345	0·00345
3	80	800	8000	0·8	0·08	0·008
4	7	70	700	0·07	0·007	0·0007
5	257	2570	25700	2·57	0·257	0·0257
6	780	7800	78000	7·8	0·78	0·078
7	7·3	73	730	0·073	0·0073	0·00073
8	456·7	4567	45670	4·567	0·4567	0·04567
9	17·85	178·5	1785	0·1785	0·01785	0·001785
10	8·34	83·4	834	0·0834	0·00834	0·000834

answer section / page 189

Exercise 7.6

	a	b	c	d	e	f	g	h	i
1	$\frac{1}{2}$	$\frac{1}{4}$	$\frac{1}{2}$	$\frac{3}{4}$	$\frac{1}{3}$	$\frac{2}{3}$	$\frac{1}{3}$	$\frac{3}{4}$	$\frac{1}{4}$
2	$\frac{1}{2}$	$\frac{1}{4}$	$\frac{1}{3}$	$\frac{2}{3}$	$\frac{1}{3}$	$\frac{1}{4}$	$\frac{1}{4}$	$\frac{3}{4}$	$\frac{1}{3}$
3	$\frac{1}{2}$	$\frac{3}{4}$	$\frac{1}{2}$	$\frac{3}{4}$	$\frac{2}{3}$	$\frac{2}{3}$	$\frac{2}{3}$	$\frac{3}{4}$	$\frac{2}{3}$

	4	5	6	7	8	9
Boys	$\frac{3}{12}=\frac{1}{4}$	$\frac{6}{8}=\frac{3}{4}$	$\frac{2}{6}=\frac{1}{3}$	$\frac{12}{16}=\frac{3}{4}$	$\frac{5}{15}=\frac{1}{3}$	$\frac{18}{24}=\frac{3}{4}$
Girls	$\frac{9}{12}=\frac{3}{4}$	$\frac{2}{8}=\frac{1}{4}$	$\frac{4}{6}=\frac{2}{3}$	$\frac{4}{16}=\frac{1}{4}$	$\frac{10}{15}=\frac{2}{3}$	$\frac{6}{24}=\frac{1}{4}$

	a	b	c	d
10	$\frac{12}{24}=\frac{1}{2}$	$\frac{8}{24}=\frac{1}{3}$	$\frac{4}{24}=\frac{1}{6}$	—
11	$\frac{6}{36}=\frac{1}{6}$	$\frac{12}{36}=\frac{1}{3}$	$\frac{15}{36}=\frac{5}{12}$	$\frac{3}{36}=\frac{1}{12}$
12	$\frac{32}{48}=\frac{2}{3}$	$\frac{16}{48}=\frac{1}{3}$	—	—
13	$\frac{52}{64}=\frac{13}{16}$	$\frac{12}{64}=\frac{3}{16}$	—	—
14	$\frac{10}{35}=\frac{2}{7}$	$\frac{25}{35}=\frac{5}{7}$	—	—

Exercise 7.7

	a	b	c	d	e
1	0·5	0·2	0·25	0·4	0·75
2	0·4285	0·2857	0·3333	0·1428	0·5
3	1·5	0·7777	2·25	1·444	2·2
4	0·4814	2·333	0·7714	2·4	0·9310
5	0·04615	0·1666	0·5945	0·4942	0·6587
6	0·6666	0·5	0·6	1·75	0·8
7	0·4	0·75	0·5714	1·4	1·090

	f	g	h	i	j
1	0·3333	0·6	0·8	0·6666	0·1666
2	0·7142	0·6666	0·5714	0·8333	0·8571
3	1·75	1·8	0·5454	0·2142	0·4545
4	0·5937	0·5156	0·6428	0·9101	0·4615
5	0·1694	0·2935	0·003348	0·5216	1·029
6	0·2	0·25	0·3333	0·5	2·666
7	0·5454	0·2857	0·8888	2·25	0·5555

Exercise 7.8

The examples up to 4(b) come out exactly. Most of the remainder do not.

	a	b	c	d	e
1	9	12	6	18	21
2	4	6	10	16	24
3	4	5	8	10	22
4	10	15	17·85	21·42	32·85
5	20·22	43·55	76·22	22·55	65·33
6	19·95	48·04	68·00	46·56	136·0
7	£0·75	£0·95	£2·737	£0·518	
8	£0·90	£1·34	£0·194	£9·968	
9	£11·416	£22·946	£0·468	£11·956	

10 6
11 32
12 a £4·50; **b** £4·65; **c** £8·40; **d** £2·698; **e** £9·168
13 a £2·75; **b** £9·46; **c** £0·88; **d** £3·79½; **e** £27·28;
14 a £8, £20; **b** £4·40, £11; **c** £12·60, £31·50;
  d £7·428, £18·571; **e** £7·017, £17·542

unit 8

The work on negative co-ordinates may be omitted as only the work in Exercise 8.1 is required in the remainder of the book. (Except for Exercise 22.3.)

Castle (1, 3), Farm (3, 6), Bridge (4, 2), Junction (1, 6), House (0, 2), Station (5, 0), Church (5, 5),
Road crossings (5, 3), (5, 4)

Exercise 8.1

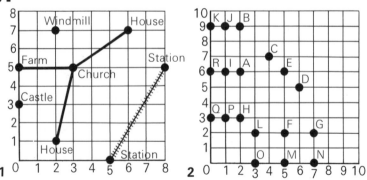

3 M (6, 5), N (1, 2), O (3, 4), P (2, 3), Q (4, 4), R (2, 2), S (1, 1), T (0, 2), U (2, 0), V (4, 2), W (6, 1), X (4, 0), Y (1, 4), Z (0, 4).

Church (1, 1), Station (1, ⁻4), Bridge (0, ⁻2),
Houses (1, 2), (⁻4, ⁻2), Farm (⁻1, 2),
Road crossings (1, 0), (1, ⁻1).

… answer section/page 191

Exercise 8.2

1 A (3, ⁻2), B (2, ⁻3), C (⁻2, 3), D (⁻2, 2), E (2, ⁻2),
F (3, ⁻1), G (⁻3, 1), H (⁻3, 3), I (3, 0), J (3, 2),
K (⁻1, 0), L (0, ⁻3), M (⁻3, ⁻3), N (⁻2, ⁻1),
O (0, 0), P (⁻2, ⁻3), Q (⁻1, 1), R (⁻3, ⁻2), S (2, 1),
T (3, ⁻3), U (⁻2, ⁻2), V (⁻3, 0), W (0, 3), X (1, ⁻3),
Y (2, 3), Z (⁻2, ⁻4).

2

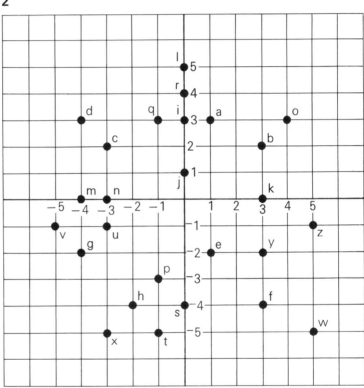

unit 9

Exercise 9.1

1 E, 10; I, 9; H, 11. **a** H, **b** I.
2 A, 20 sq cm B, 21 sq cm; C, 24 sq cm; D, 22 sq cm;
E, 72 sq cm.
3 **a** 15 sq cm; **b** 16 sq cm; **c** 30 sq cm; **d** 63 sq cm;
e 120 sq cm; **f** 300 sq m; **g** 10 sq cm; **h** 17·5 sq m;
i 2·25 sq m; **j** 5·12 sq cm; **k** 10·73 sq m.
4 **a** 20 sq m; **b** £60.
5 **a** 14 sq m; **b** 7 bags; **c** £3·50.
6 **a** 24 sq m; **b** 6 tins; **c** £1·62.
7 **a** 18 sq m; **b** £54.
8 **a** 6 sq m; **b** 4 bags; **c** £2.

Exercise 9.2

1. **a** 6 sq cm; **b** 10 sq cm; **c** 24 sq cm; **d** 30 sq cm; **e** 32 sq cm; **f** 31·5 sq cm; **g** 7·5 sq cm; **h** 49·5 sq cm; **i** 12 sq cm; **j** 14 sq cm; **k** 20 sq cm.
2. **a** 4 sq cm; **b** 24 sq cm; **c** 15 sq m; **d** 28 sq cm; **e** 31·5 sq m; **f** 84 sq cm; **g** 93·5 sq cm; **h** 3 sq cm; **i** 4·32 sq cm; **j** 12·455 sq cm.
3. **a** 40 sq m; **b** £3·60.
4. **a** 24 sq m; **b** £2·16.
5. **a** 21 sq m; **b** £1·89.
6. **a** 31·5 sq m; **b** £2·83$\frac{1}{2}$.
7. **a** Base 6 cm, height 3·3 cm; Area 9·9 sq cm.
   b Base 5 cm, height 4·0 cm; Area 10·0 sq cm.
   c Base 4 cm, height 5·0 cm; Area 10·0 sq cm.
   d Yes, The area correct to 3 significant figures is 9·92 sq cm.
8. **a** 6 sq m; **b** 1500 g = 1·5 kg; **c** £1·09$\frac{1}{2}$.

Exercise 9.3

Some pupils may have difficulty in deciding the measurements of the component rectangles in questions 1(a) to 1(e). If this is so they should first draw them on squared paper, and if necessary count the squares.

1. **a** 26 sq cm; **b** 24 sq cm; **c** 24 sq cm; **d** 22 sq cm; **e** 56 sq cm.
2. **b** 36 sq m.
3. 17·84 sq m.
4. **a** 16 sq m; **b** 32 sq m; **c** 67 sq m; **d** 9·2 sq cm.
5. **b** 3800 sq cm = 0·38 sq m.

Exercise 9.4

It may be helpful to illustrate the concept of volume by demonstrating and supplementing the examples in the text by actually using wooden blocks.

1. **a** 48 cubic cm; **b** 175 cubic cm; **c** 72 cubic cm; **d** 160 cubic cm; **e** 693 cubic cm; **f** 189 cubic cm.
2. **a** 144 cubic cm; **b** 70 cubic cm; **c** 20·25 cubic m; **d** 279·1 cubic cm; **e** 7429 cubic cm.
3. **a** 108 cubic m, 18; **b** 240 cubic m, 40; **c** 384 cubic m, 64; **d** 600 cubic m, 100; **e** 459 cubic m, 76 (76·5).
4. **a** 196 000 cubic cm; **b** 125 cubic cm; **c** 1568.
5. 2500.

Exercise 9.5

1 **a** 128 sq cm; **b** 122 sq cm; **c** 94 sq cm.
2 **a** 180 sq cm; **b** 130 sq cm; **c** 166 sq m;
  d 1198 sq cm; **e** 2302 sq cm.
3 **a** (i) 3·28 sq m; (ii) 32 800 sq cm; **b** 6 tins (5·47);
  c 72p.

unit 10

Exercise 10.1

1 $D = 4L$
2 $D = 4a$
3 $D = 2a + 2b$
4 $D = 2l + 2w$
5 $D = 3s$
6 $D = 3b$
7 $D = x + 2y$
8 $D = b + 2s$
9 $D = 2a + 3b$
10 $D = b + 2s + t$
11 $D = 2x + 6y$
12 $D = 4s$
13 $D = 2a + b$
14 $D = x + y + z$
15 $D = e + f + g$
16 $D = x + w + y + z$
17 $D = 2a + 2b + 2c$
18 $D = 6s$
19 $D = 2b + 2s$
20 $W = 4L + 8S$
21 $W = 6a$
22 $W = 4H + 4L + 4S$
23 $W = 4a + 4b + 4c$
24 $W = 4B + 4E$

Exercise 10.2

1 $N = B + G, B = N - G$
2 $M = T - W, T = M + W$
3 $N = D + C, N = C + D$
4 $B = T - W, T = B + W$
5 $T = C + X, X = T - C$
6 $T = PY, T = YP$
7 $T = NP, N = T \div P$
8 $W = BG, B = W \div G$
9 $L = RC, R = L \div C$
10 $T = AB, A = T \div B$
11 **a** $T = B + S$; **b** $S = T - B$; **c** $B = T - S$
12 **a** $T = B + C$; **b** $C = T - B$; **c** $B = T - C$
13 **a** $T = BM$; **b** $B = T \div M$; **c** $M = T \div B$
14 **a** $R = LN$; **b** $L = R \div N$; **c** $N = R \div L$
15 $X + Y + Z + W$
16 $P - Y$
17 $P - X - Y$
18 $PX + RY$
19 **a** hp; **b** $hp \div 100$
20 **a** $bx + cy$; **b** $P - bx - cy$

Exercise 10.3

	a	b	c	d
1	$4x$	$2x + 3y$	$2x + 2y$	$5y$
2	$2a + b$	$2a + 3b$	$6a$	$18a$
3	AB	LW	PQ	$4B$
4	$4a$	$2a$	$4b$	b
5	0	$11b$	c	$5a$

'N' means 'not possible to simplify'.

6	$9a+9b$	$3m+7n$	$3x+4y$	$7b$
7	$8a$	$21a$	$6ab$	abc
8	$15ab$	$40gf$	$18qp$	$12pq$
9	$3a$	$4b$	$4q$	$2p$
10	$4a$	$4b$	$5q$	$5p$
11	10	16	18	6
12	$2ab$	$12q$	$2f$	3
13	$5ab$	$16p$	$5e$	2
14	$7x$	N	$2x+y$	N
15	$a+4b$	N	b	N
16	$6ab$	$12a$	$10e$	$8rf$
17	2	N	$\dfrac{b}{2c}$	$\dfrac{9b}{4a}$

unit 11

Exercise 11.1

	a	b	c	d	e
1	32	25	7	0	72
2	13	4	1	8	9
3	10	15	6	12	14
4	3	2	⁻3	3	5
5	4	⁻5	⁻17	0	7
6	41	15	19	4	29
7	1	9	1	⁻4	7
8	40	15	24	5	0
9	120	60	0	21	120
10	4	8	2	4	5
11	2·2	4·333	1·625	5·333	2·571

Exercise 11.2

	a	b	c	d	e	f
1	24 sq m	72 sq m	90 sq m	168 sq m	108 sq m	—
2	192 g	384 g	1200 g	672 g	1440 g	1296 g
3	70 min	50 min	90 min	110 min	150 min	170 min
	g 190 min	h 310 min	i 330 min	j 530 min		
4	49 sq m	64 sq m	72 sq m	48 sq m	58·5 sq m	—
5	£30	£60	£120	£175	£360	—
6	120 km	180 km	300 km	540 km	720 km	—
7	£43	£67	£82	£112	£223	—
8	24 sq m	32 sq m	66 sq m	104 sq m	90 sq m	144 sq m
9	£7	£8	£5	£9	£14	£4·50
10	10 sq m	18 sq m	28 sq m	40 sq m	66·5 sq m	—

unit 12

The examples in Exercises 1 and 2 all come out exactly, and are aimed at teaching the idea of a percentage. The examples can all be done by inspection. More difficult examples which require calculation are in Exercises 3 and 4.

answer section/page 195

Exercise 12.1

	a	b	c	d
1	6%	4%	3%	4%
2	6%	28%	25%	30%
3	8%	56%	30%	8%
4	50%	25%	10%	10%
5	20%	34%	34%	80%
6	8·5%	5%	14%	75%
7	70%	60%	40%	55%
8	75%	72%	2A	—
9	26%	20%	Safedrive	—
10	60%	40%	100%	—
11	80%	20%	100%	—
12	30%	—	—	—
13	20%	—	—	—
14	28%	—	—	—
15	30%	—	—	—
16	12%	10%	Extralux	—

Exercise 12.2

	a	b	c	d	e	f
1	20	40	10	5	4	2
2	40	80	20	10	8	4
3	4	16	8	24	32	2
4	15	6	3	60	90	150
5	9	45	90	135	180	225
6	40	20	16	8	4	32
7	4	2	1	8	—	—
8	20	10	5	40	60	—
9	£10	£5	£2	£20	—	—
10	40 kg	20 kg	10 kg	8 kg	4 kg	—
	60 kg	30 kg	15 kg	12 kg	6 kg	—
11	£4	£2	£1	£96	£48	£24

Exercise 12.3

	a	b	c	d	e
1	80·00%	66·66%	57·14%	50·00%	44·44%
2	83·33%	71·42%	62·50%	55·55%	45·45%
3	85·71%	75·00%	66·66%	54·55%	46·15%
4	64·28%	75·00%	40·90%	20·93%	15·78%
5	66·07%	54·41%	39·36%	49·33%	23·41%
6	77·78%	75·00%	1972	—	—
7	87·5%	77·77%	No	—	—
8	22·22%	20·00%	3G	—	—
9	29·16%	28·57%	Everlast, but only slightly so.		
10	Yes, 5·769%	No, 4%	No, 1·626%	—	

page 196/answer section

11 3·846%   4·838%   —   —   —
12 1·2%   1·066%   0·6666% No, error slightly
                                 over 1% in **a** and **b**
13 £90   £12   15·38%   —   —
14 £1080   £97   9·867%   —   —

Exercise 12.4

	a	b	c	d	e
1	£0·35	£0·45	£0·30	£0·15	£0·40
2	£0·69	£1·15	£1·61	£1·84	£2·07
3	2·24 kg	3·29 kg	4·13 kg	2·03 kg	5·88 kg
4	£0·38½	£0·49½	£0·22	£0·11	£0·27½
5	0·62 m	0·6975 m	1·24 m	2·7125 m	0·651 m
6	7·559p	11·31p	£0·5522	1·819p	0·7933p
7	£2·04	£1·44	21·36p	—	—
8	£24	£54·72	£237·12	£22·61	85·44p
9	9	15	679	2134	—
10	16·1 kg	3·45 kg	55·2 kg	55·2t	—
	31·5 kg	6·75 kg	108 kg	108t	—
	22·4 kg	4·8 kg	76·8 kg	76·8t	—
11	£5·60	£20·80	£37·04	£280	£299·44
12	£4·34	£18·60	£26·815	£193·75	£182·667

unit 13

Exercise 13.1

1 **a** d^4; **b** $k \times k \times k$
2 **a** h^3; **b** $h \times h$
3 **a** k^2; **b** $d \times d \times d \times d \times d$
4 **a** g^5; **b** $e \times e \times e$
5 **a** w^3; **b** $u \times u \times u \times u$
6 **a** c^4; **b** $s \times s$
7 **a** r^2; **b** $w \times w \times w \times w \times w$
8 **a** y^5; **b** p
9 **a** r, (r^1 is correct, but not necessary);
   b $w \times w \times w \times w \times w$

Exercise 13.2

1 **a** 9 sq cm; **b** 49 sq cm; **c** 81 sq cm; **d** 169 sq cm.
2 **a** 27 cubic cm; **b** 64 cubic cm; **c** 343 cubic cm;
   d 729 cubic cm.

answer section/page 197

	a	b	c	d
3	2	4	8	16
4	3	9	27	81
5	5	25	125	625
6	5	13	25	41
7	65	61	113	458
8	9	7	5	3
9	33	11	15	135
10	32	141	6	287
11	25	49	4	1
12	8	1	125	343
13	5	29	74	53
14	21	24	45	48
15	12	2	75	196
16	11	92	25	8

unit 14

Exercise 14.1

A number by itself indicates that the number is prime.

1. **a** 3; **b** 3×3; **c** 5; **d** 2×8; **e** 2×5.
2. **a** 2×2; **b** 2×7; **c** $2 \times 6 = 3 \times 4$; **d** 17; **e** 3×5.
3. **a** 19; **b** 3×7; **c** $2 \times 10 = 4 \times 5$;
   d $2 \times 15 = 3 \times 10 = 5 \times 6$; **e** $2 \times 12 = 3 \times 8 = 4 \times 6$.
4. **a** 2×11; **b** 5×5; **c** 23; **d** 3×9; **e** 3×11.
5. **a** 2×13; **b** 31; **c** $2 \times 14 = 4 \times 7$; **d** $2 \times 16 = 4 \times 8$;
   e 29.
6. **a** 43; **b** $2 \times 18 = 3 \times 12 = 4 \times 9 = 6 \times 6$; **c** 3×13;
   d 2×17; **e** 37.
7. **a** 5×7; **b** $2 \times 20 = 4 \times 10 = 5 \times 8$; **c** 2×19; **d** 41;
   e 3×11.
8. **a** $2 \times 21 = 3 \times 14 = 6 \times 7$; **b** $2 \times 32 = 4 \times 16 = 8 \times 8$;
   c $2 \times 36 = 3 \times 24 = 4 \times 18 = 6 \times 12 = 8 \times 9$; **d** 97; **e** 101.
9. **a** $2 \times 30 = 3 \times 20 = 4 \times 15 = 5 \times 12 = 6 \times 10$; **b** 107;
   c $3 \times 27 = 9 \times 9$; **d** 7×13; **e** 3×31.
10. **a** $2 \times 50 = 4 \times 25 = 5 \times 20 = 10 \times 10$; **b** 67;
    c $2 \times 72 = 3 \times 48 = 4 \times 36 = 6 \times 24 = 8 \times 18 = 9 \times 16 = 12 \times 12$;
    d 11×11; **e** 113.

Exercise 14.2

1. **a** $2 \times 2, \{1, 2, 4\}$; **b** $2 \times 3, \{1, 2, 3, 6\}$; **c** $3 \times 3, \{1, 3, 9\}$;
   d $2 \times 2 \times 2 \times 2, \{1, 2, 4, 8, 16\}$;
   e $2 \times 2 \times 5, \{1, 2, 4, 5, 10, 20\}$.
2. **a** $3 \times 7, \{1, 3, 7, 21\}$; **b** $2 \times 11, \{1, 2, 11, 22\}$;
   c $2 \times 2 \times 2 \times 3, \{1, 2, 3, 4, 6, 8, 12, 24\}$;
   d $5 \times 5, \{1, 5, 25\}$; **e** $2 \times 2 \times 7, \{1, 2, 4, 7, 14, 28\}$.

3 **a** $2\times3\times5$, {1, 2, 3, 5, 6, 10, 15, 30};
 b $2\times2\times2\times2\times2$, {1, 2, 4, 8, 16, 32};
 c 3×11, {1, 3, 11, 33}; **d** 2×17, {1, 2, 17, 34};
 e 5×7, {1, 5, 7, 35}.
4 **a** $2\times2\times3\times3$, {1, 2, 3, 4, 6, 9, 12, 18, 36};
 b $2\times2\times2\times5$, {1, 2, 4, 5, 8, 10, 20, 40};
 c $3\times3\times5$, {1, 3, 5, 9, 15, 45};
 d $2\times2\times2\times2\times3$, {1, 2, 3, 4, 6, 8, 12, 16, 24, 48};
 e $2\times2\times2\times3\times3$, {1, 2, 3, 4, 6, 8, 9, 12, 18, 24, 36, 72}.
5 **a** $2\times2\times2\times2\times2\times2$, {1, 2, 4, 8, 16, 32, 64};
 b $2\times5\times7$, {1, 2, 5, 7, 10, 14, 35, 70};
 c $2\times2\times19$, {1, 2, 4, 19, 38, 76};
 d $2\times2\times5\times5$, {1, 2, 4, 5, 10, 20, 25, 50, 100};
 e $2\times2\times2\times2\times3\times3$, {1, 2, 3, 4, 6, 8, 9, 12, 16, 18, 24, 36, 48, 72, 144}.

1 **a** 2^2; **c** 3^2; **d** 2^4; **e** $2^2\times5$.
2 **c** $2^3\times3$; **d** 5^2; **e** $2^2\times7$.
3 **b** 2^5.
4 **a** $2^2\times3^2$; **b** $2^3\times5$; **c** $3^2\times5$; **d** $2^4\times3$; **e** $2^3\times3^2$.
5 **a** 2^6; **c** $2^2\times19$; **d** $2^2\times5^2$; **e** $2^4\times3^2$.

Exercise 14.3

1 53, 59, 61, 67, 71, 73, 79, 83, 89, 97.
2 101, 103, 107, 109, 113, 127, 131, 137, 139, 149.
3 503, 509, 521, 523, 541, 547.

Exercise 14.4

	a	b	c	d	e
1	Prime	3	Prime	2	Prime
2	Prime	Prime	5	Prime	3
3	Prime	7	Prime	3	Prime
4	13	Prime	Prime	13	13
5	Prime	17	23	Prime	29

Exercise 15.1

1 **a** Brown, **b** 7; **c** 12; **d** 32.
2 **a** Soccer; **b** 7; **c** 9.
3 **a** 8; **b** 25; **c** 34.
4 **a** 130; **b** 110; **c** 100; **d** 90; **e** 70.
5 **a** 50p; **b** 55p; **c** $37\frac{1}{2}$p.

Exercise 15.2

1 **a** 7; **b** $10+5=15$; **c** 30.

3 a 1; **b** 3; **c** Wednesday; **d** Monday; **e** 18; **f** It need not be the same; for example 2 absences can come from Brown being away on Monday and Tuesday, or Brown being away only on Monday, and Smith on Tuesday.
5 a Tuesday and Thursday; **b** Saturday; **c** Tuesday with Thursday, Wednesday with Friday, Monday with Sunday; **d** 2; **e** 18.
7 2, 3, 1, 1, 5, 2, 2, 2, 4, 3.
8 100, 85, 65, 90, 75, 85, 90, 75, 110, 100, 185, 135.

Exercise 15.3

The first 8 examples come out exactly, some of the remainder do not.

1 a 6; **b** 8; **c** 3; **d** 5; **e** 56; **f** 70; **g** 47; **h** 56.
2 6.   **3** $4\frac{1}{2}$p.   **4** 15 °C.   **5** 1·52 m.   **6** £105.   **7** 61%.
8 84 kg, Team A.
9 a 11·44; **b** 29·14; **c** 45·87; **d** 4·333; **e** 12·77.
10 £141.
11 a 7; **b** 6·769; **c** The girls; **d** 6·875;   **12** 91 grammes.

Exercise 15.4

Questions 4, 5, and 6 require tracing paper.

1 a £1; **b** £5; **c** £2; **d** £3; **e** £2; **f** £3.
2 a £8; **b** £6; **c** £0; **d** £2; **e** £12; **f** £4.
3 a £24; **b** £12; **c** £12; **d** £64.
7 a £12; **b** £3; **c** £24.
8 Mary 40p; Joan 30p; John 50p.

unit 16

An alternative to doing the entire Unit is to do Exercise 16.2 examples 1(a) and 1(b) to 10(a) and 10(b), and then to do Exercise 16.4 examples 1 to 15, all of which come out exactly.

Exercise 16.1

	a	b	c	d
1	$g=6$	$h=6$	$b=14$	$y=10$
2	$a=6$	$b=4$	$d=16$	$x=16$
3	$c=13$	$x=4$	$y=21$	$f=28$
4	$d=11$	$y=13$	$p=69$	$g=63$
5	$e=12$	$s=33$	$h=82$	$s=149$
6	$e=49$	$d=8$	$f=203$	$g=690$
7	$d=5·6$	$f=1·8$	$h=10·1$	$s=8·2$
8	$f=24$	$g=55$	$d=3\frac{1}{4}$	$w=2\frac{2}{3}$
9	$r=0$	$g=15$	$g=8$	$x=28$
10	$s=5$	$k=7$	$r=5$	$d=15$

Exercise 16.2

	a	b	c	d
1	$b=4$	$b=3$	$g=12$	$k=21$
2	$a=3$	$b=2$	$a=6$	$v=12$
3	$b=9$	$c=3$	$b=20$	$w=35$
4	$f=9$	$s=8$	$c=30$	$x=30$
5	$d=10$	$h=9$	$d=56$	$y=16$
6	$r=3$	$h=6$	$e=63$	$z=72$
7	$x=62$	$x=109$	$h=31.8$	$g=4.68$
8	$v=5.75$	$f=3.375$	$g=8.6$	$h=4$
9	$c=2.4$	$b=0.9$	$u=42.6$	$g=2.8$
10	$y=1.567$	$c=0.3714$	$f=17$	$k=1.19$
11	$m=33$	$w=6.4$	$u=6.08$	$e=0.21$

Exercise 16.3

	a	b	c	d
1	$x=4$	$b=5$	$y=6$	$h=48$
2	$s=4$	$z=5$	$y=6$	$z=14$
3	$x=4$	$v=6$	$x=5$	$y=21$
4	$z=2$	$a=5$	$e=4$	$f=54$
5	$d=3$	$f=6$	$w=0$	$v=27$
6	$d=0$	$r=1$	$h=18$	$k=4$
7	$q=2.5$	$v=3.5$	$g=4$	$h=160$
8	$a=\;^-3$	$a=\;^-5$	$d=\;^-28$	$c=468$
9	$w=\;^-0.5714$	$c=\;^-4.5$	$g=2$	$g=98$
10	$b=4.333$	$r=2.25$	$k=\;^-198$	$k=364$

Exercise 16.4

1 6 m   **2** 4 m   **3** 2 m   **4** 4 m   **5** 3 m
6 2 m   **7** 2 m   **8** 3 m   **9** 3 m   **10** 4 m

In the next five examples the weights refer to sugar, butter, and flour respectively.

11 3 kg, 6 kg, 15 kg   **16** (1) 6·857 m   **17** (6) 2·333 m
12 2 kg, 4 kg, 8 kg   (2) 4·8 m   (7) 1·25 m
13 2 kg, 6 kg, 10 kg   (3) 2·571 m   (8) 3·333 m
14 80 g, 120 g, 200 g   (4) 10·66 m   (9) 3·125 m
15 100 g, 150 g, 200 g   (5) 2·333 m   (10) 4·2 m

In the next five examples the weights refer to sugar, butter, and flour respectively.

18 (11) 87·5 g, 175 g, 437·5 g
    (12) 114·2 g, 228·5 g, 457·1 g
    (13) 1·111 kg, 3·333 kg, 5·555 kg
    (14) 70 g, 105 g, 175 g
    (15) 111·1 g, 166·6 g, 222·2 g

answer section/page 201

Exercise 16.5

1. 13·88 m
2. 15·16 m
3. 6
4. 5
5. 3 kg
6. 6 kg
7. 50 °F
8. 68 °F
9. 5 °C
10. −5 °C
11. 5 kg
12. 3 kg
13. 7 kg
14. £110, £220, £330
15. £120, £360
16. 20 kg, 40 kg, 80 kg
17. 50 g, 100 g, 250 g
18. 400 kg, 600 kg, 800 kg
19. 11
20. 40
21. 88
22. 13·2 m
23. 2·5
24. 20 years

unit 17

Exercise 17.1

These examples all come out exactly by the method shown.

	a	b	c	d	e
1	£6, £3	£6, £2	£4, £8	£9, £3	£4, £6
2	£12, £8	£3, £3	£12, £3	£6, £8	£2, £8
3	£12, £9	£4, £10	£10, £6	£8, £10	£9, £15
4	£15, £6	£15, £12	£10, £2	£12, £2	£18, £15
5	£35, £42	£70, £30	£56, £35	£90, £110	£42, £34
6	£2, £3	£6, £9	£10, £15	£30, £45	—
7	£3, £7	£6, £14	£9, £21	—	—
8	£5, £3	£15, £9	£20, £12	£30, £18	—
9	8 kg, 20 kg	10 kg, 25 kg	20 kg, 50 kg	—	—
10	4 kg, 1 kg	8 kg, 2 kg	16 kg, 4 kg	32 kg, 8 kg	—

Exercise 17.2

Most of these examples do not come out exactly, and it is recommended that some of Exercise 17.1 should be done by the equation method before this exercise is done.

	a	b	c
1	£12, £4	£6·40, £9·60	£12·444, £3·555
2	£13·80, £9·20	£14·375, £8·625	£11·50, £11·50
3	£26·285, £19·714	£13·142, £32·857	£28·75, £17·25
4	£16·835, £6·734	£13·094, £10·475	£19·641, £3·928
5	£6·786, £8·143	£10·451, £4·479	£9·187, £5·742
6	£3·636, £6·363	£1·818, £3·181	£9·309, £16·290
7	2·857 kg, 7·142 kg	6·857 kg, 17·14 kg	9·142 kg, 22·85 kg
8	1·428 kg, 8·571 kg	0·8571 kg, 5·142 kg	0·2857 kg, 1·714 kg

page 202/answer section

	d	e
1	£7·111, £8·888	£8·727, £7·272
2	£6·571, £16·428	£12·545, £10·454
3	£20·444, £25·555	£17·25, £28·75
4	£20·202, £13·367	£12·856, £10·713
5	£6·718, £18·211	£8·250, £6·679
6	£1·189, £2·080	—
7	21·42 kg, 53·57 kg	—
8	71·42 g, 428·5 g	—

Exercise 17.3

	a	b	c
1	3:4	2:3	5:2
2	3:4	2:3	5:2
3	1:2	2:5	7:3
4	1:3	2:3	7:3
5	3:2	1:3	1:4
6	2:3	£20, £30	£30, £45
7	4:3	16 kg, 12 kg	20 kg, 15 kg
8	5:3	£10, £6	£12·50, £7·50
9	£30, £50	£56, £24	£30, £50
10	£40, £60	£44, £66	£310·40, £465·60

	d	e
1	2:1	2:3
2	3:5	5:3
3	3:5	8:5
4	3:1	4:1
5	3:1	3:5
6	£12·80, £19·20	—
7	48·57 kg, 36·42 kg	—
8	£14·375, £8·625	—
9	£45·714, £34·285	—
10	£13 826·80, £20 740·20	£138 657·20, £207 985·80

Exercise 17.4

	a	b	c
1	£1, £2, £3	£2, £3, £4	£5, £4, £2
2	4p, 6p, 8p	10p, 8p, 4p	4p, 6p, 6p
3	15 m, 12 m, 6 m	6 m, 9 m, 9 m	4 m, 8 m, 12 m
4	8 kg, 12 kg, 12 kg	5 kg, 10 kg, 15 kg	10 kg, 15 kg, 20 kg
5	£6, £8, £2	£9, £12, £3	£12, £16, £4
6	£2, £3, £5	£4, £6, £10	£6, £9, £15

answer section/page 203

	d	e
1	£2, £3, £3	£2, £4, £6
2	3p, 6p, 9p	6p, 9p, 12p
3	8 m, 12 m, 16 m	20 m, 16 m, 8 m
4	25 kg, 20 kg, 10 kg	10 kg, 15 kg, 15 kg
5	£15, £20, £5	£75, £100, £25
6	£8, £12, £20	£1, £1·50, £2·50

7
a £0·888, £2·666, £4·444
b £1·777, £5·333, £8·888
c £2·666, £7·999, £13·333
d £4·444, £13·333, £22·222
e £22·222, £66·666, £111·11

8
£4·166, £8·333, £12·499
£5·555, £8·333, £11·111
£11·363, £9·090, £4·545
£6·25, £9·375, £9·375
£9·375, £12·500, £3·125

9
a 7·2 kg, 10·8 kg, 18 kg
b 4 kg, 12 kg, 20 kg
c 12 kg, 20 kg, 4 kg
d 18 kg, 4·5 kg, 13·5 kg
e 6·545 kg, 13·09 kg, 16·36 kg

10
18·66 m, 13·99 m, 9·333 m
11·2 m, 14 m, 16·8 m
12 m, 14 m, 16 m
18·9 m, 4·2 m, 18·9 m
10·5 m, 13·5 m, 18 m

11
a 7·142 kg, 14·28 kg, 28·57 kg
b 6·142 kg, 12·28 kg, 24·57 kg
c 1·428t, 2·857t, 5·714t
d 0·7142t, 1·428t, 2·857t
e 0·142t, 0·2857t, 0·5714t

12
28·57 g, 114·2 g, 57·14 g, 1 egg
57·14 g, 228·5 g, 114·2 g, 2 eggs
114·2 g, 457·1 g, 228·5 g, 4 eggs
142·8 g, 571·4 g, 285·7 g, 5 eggs
714·2 g, 2857 g, 1428 g, 25 eggs

	a	b	c	d	e
13	1:2:3	3:4:1	2:3:4	5:4:2	2:3:5
14	5:4:2	1:2:3	2:3:4	1:2:1	3:4:6
15	5:7:9	1:2:3	4:5:6	1:2:3	2:3:6

16
a 1:2:3
b £2, £4, £6
c £3·333, £6·666, £9·999
d £16·666, £33·333, £50
e £33 333·333, £66 666·666, £100 000

17
£25·999, £17·333, £34·666
£20·80, £26, £31·20
£13, £26, £39
£14·625, £24·375, £39
—

18
a 5:3:2
b 90 kg, 54 kg, 36 kg
c 100 kg, 60 kg, 40 kg
d 25 g, 15 g, 10 g
e 125t, 75t, 50t

19
4:2:3
£44·444, £22·222, £33·333
£68·888, £34·444, £51·666
£55·062, £27·531, £41·296

unit 18

The pupils may take the value of π as either 3, 3·1, or 3·14.

Exercise 18.1

	π=3	π=3·1	π=3·14		π=3	π=3·1	π=3·14
1	6 cm	6·2 cm	6·28 cm	9	33 cm	34·1 cm	34·54 cm
2	9 cm	9·3 cm	9·42 cm	10	36 cm	37·2 cm	37·68 cm
3	12 cm	12·4 cm	12·56 cm	11	6 m, 18 m	18·6 m	18·84 m
4	15 cm	15·5 cm	15·7 cm	12	12 m	12·4 m	12·56 m
5	18 cm	18·6 cm	18·84 cm	13	60 m	62 m	62·8 m
6	24 cm	24·8 cm	25·12 cm	14	36 m	37·2 m	37·68 m
7	27 cm	27·9 cm	28·26 cm	15	75 m	77·5 m	78·5 m
8	30 cm	31 cm	31·4 cm	16	90 m	93 m	94·2 m

Exercise 18.2

	π=3	π=3·1	π=3·14
1	12 sq m	12·4 sq m	12·56 sq m
2	27 sq m	27·9 sq m	28·26 sq m
3	48 sq m	49·6 sq m	50·24 sq m
4	108 sq m	111·6 sq m	113·04 sq m
5	147 sq m	151·9 sq m	153·86 sq m
6	192 sq m	198·4 sq m	200·96 sq m
7	243 sq m	251·1 sq m	254·34 sq m
8	300 sq m	310 sq m	314 sq m
9	363 sq m	375·1 sq m	379·94 sq m
10	432 sq m	446·4 sq m	452·16 sq m
11	5 m, 75 sq m	77·5 sq m	78·5 sq m
12	507 sq m	523·9 sq m	530·66 sq m
13	147 sq m	151·9 sq m	153·86 sq m
14	675 sq m	697·5 sq m	706·5 sq m
15	300 sq m	310 sq m	314 sq m
16	1200 sq m	1240 sq m	1256 sq m

Exercise 18.3

π=3

1 **a** 24 m; **b** £1·44
2 **a** 27 sq m; **b** £1·89
3 **a** 6 m; **b** 42 m
4 **a** 3 m; **b** 27 sq m; **c** 243 kg
5 **a** 18 m; **b** £54
6 **a** 243 sq m; **b** £1265
7 67 500
8 **a** 30 m; **b** 15; **c** 90 m
9 **a** 9 m; **b** 18 m; **c** 1080 m
10 **a** 300 sq cm; **b** 3600 sq cm; **c** 9
11 **a** 942 cm; **b** £16·296

answer section/page 205

$\pi = 3.1$

1. **a** 24·8 m; **b** £1·488
2. **a** 27·9 sq m; **b** £1·953
3. **a** 6·2 m; **b** 43·4 m
4. **a** 3 m; **b** 27·9 sq m; **c** 251·1 kg
5. **a** 18·6 m; **b** £55·80
6. **a** 251·1 sq m; **b** £1305·50
7. 69 750
8. **a** 31 m; **b** 16; **c** 93 m
9. **a** 9·3 m; **b** 18·6 m; **c** 1116 m
10. **a** 310 sq cm; **b** 3720 sq cm; **c** 10 (9.3)
11. **a** 973·4 cm; **b** £16·839

$\pi = 3.14$

1. **a** 25·12 m; **b** £1·507
2. **a** 28·26 sq m; **b** £1·978
3. **a** 6·28 m; **b** 43·96 m
4. **a** 3 m; **b** 28·26 sq m; **c** 254·34 kg
5. **a** 18·84 m; **b** £56·52
6. **a** 254·34; **b** £1321·70
7. 70 650 sq m
8. **a** 31·4 m; **b** 16; **c** 94·2 m
9. **a** 9·42 m; **b** 18·84 m; **c** 1113·04 m
10. **a** 314 sq cm; **b** 3768 sq cm; **c** 10 (9.42)
11. **a** 985·96 cm; **b** £17·057

unit 19

Exercise 19.1

1. $\frac{6}{10}$
2. $\frac{3}{10}$
3. $\frac{1}{10}$
4. $\frac{6}{10}$
5. $\frac{3}{10}$
6. $\frac{1}{10}$
7. $\frac{4}{10}$
8. $\frac{9}{10}$
9. $\frac{6}{10}$
10. $\frac{10}{10} = 1$
11. **a** $\frac{6}{10}$; **b** $\frac{4}{10}$; **c** $\frac{10}{10} = 1$
12. **a** $\frac{1}{10}$; **b** $\frac{0}{10} = 0$
13. PQ, PR, PS, QR, QS, RS
14. $\frac{3}{6}$
15. $\frac{3}{6}$
16. $\frac{1}{6}$
17. $\frac{5}{6}$
18. $\frac{5}{6}$
19. $\frac{4}{6}$
20. $\frac{6}{6} = 1$

Exercise 19.2

1. 25; 26; 52; 56; 62; 65
2. $\frac{1}{6}$
3. $\frac{1}{6}$
4. $\frac{0}{6} = 0$
5. $\frac{2}{6}$
6. $\frac{2}{6}$
7. $\frac{4}{6}$
8. $\frac{0}{6} = 0$
9. $\frac{6}{6} = 1$
10. AET; ATE; EAT; ETA; TAE; TEA

11 $\frac{1}{6}$
12 $\frac{1}{6}$
13 $\frac{2}{6}$
14 $\frac{2}{6}$
15 $\frac{3}{6}$
16 $\frac{2}{6}$
17 $\frac{4}{6}$
18 $\frac{6}{6}=1$
19 AT; AE; AR; TE; TR; ER; TA; EA; RA; ET; RT; RE
20 $\frac{1}{12}$
21 $\frac{6}{12}$
22 $\frac{2}{12}$
23 $\frac{10}{12}$
24 $\frac{8}{12}$
25 $\frac{6}{12}$
26 $\frac{2}{12}$

Exercise 19.3

All these examples give integer answers, except when numbers 6, 7, and 8 are repeated with decimal probabilities.

	a	b	c	d	e	f	g
1	4	8	10	14	24	—	—
2	2	4	5	7	12	—	—
3	6	9	12	21	42	—	—
4	8	16	20	32	52	—	—
5	6	9	12	18	60	—	—
6	6	4	8	10	20	—	—
7	4	5	8	10	100	—	—
8	6	16	18	20	68	—	—
9	$\frac{8}{20}=\frac{2}{5}$	10	18	20	30	—	—
10	$\frac{6}{24}=\frac{1}{4}$	4	8	12	10	36	—
11	$\frac{20}{70}=\frac{2}{7}$	12	14	16	68	—	—
12	$\frac{8}{36}=\frac{2}{9}$	4	6	10	16	48	74
13	$\frac{5}{45}=\frac{1}{9}$	2	4	8	9	54	—

Numbers **1** to **5** repeated with decimals give same answers as above.

6 6(5·94) 4(3·96) 8(7·92) 10(9·9) 20(19·8) — —
7 4(3·92) 5(4·9) 8(7·84) 10(9·8) 98 — —
8 6(5·94) 16(15·84) 18(17·82) 20(19·8) 67(67·32) — —

Exercise 19.4

1 THH, HTT, THT, TTH, TTT
2 a $\frac{1}{8}$; b $\frac{3}{8}$; c $\frac{3}{8}$; d $\frac{1}{8}$
3 1, 3, 3, 1
5 The bar charts in this question and in question **4** are unlikely to be exactly the same, but even with only 8 throws we can in general expect the middle two bars to be longer than the ones at the end.
6 HTHH, THHH, HHTT, HTHT, THHT, THTH, TTHH, HTTH, TTTH, TTHT, THTT, HTTT, TTTT
7 a $\frac{1}{16}$; b $\frac{4}{16}$; c $\frac{6}{16}$; d $\frac{4}{16}$; e $\frac{1}{16}$
8 2, 8, 12, 8, 2

answer section/page 207

10 The bar charts in this question and in question **9** are unlikely to be exactly the same, but we can in general expect that the middle bars will be longer than the bars at the end.

Exercise 19.5

Questions 13 to 24 give integer answers for the 72 throws, but mainly non-integer answers when repeated for 24 throws.

1 $\frac{6}{36} = \frac{1}{6}$ **7** $\frac{10}{36} = \frac{5}{18}$ **13** 12   **19** 20   **13** 4   **19** 7 (6·666)
2 $\frac{2}{36} = \frac{1}{18}$ **8** $\frac{15}{36} = \frac{5}{12}$ **14** 4   **20** 30   **14** 1 (1·333)   **20** 10
3 $\frac{1}{36}$ **9** $\frac{15}{36} = \frac{5}{12}$ **15** 2   **21** 30   **15** 1 (0·6666)   **21** 10
4 $\frac{5}{36}$ **10** $\frac{21}{36} = \frac{7}{12}$ **16** 10   **22** 42   **16** 3 (3·333)   **22** 14
5 $\frac{5}{36}$ **11** $\frac{16}{36} = \frac{4}{9}$ **17** 10   **23** 32   **17** 3 (3·333)   **23** 11 (10·66)
6 $\frac{0}{36} = 0$ **12** $\frac{10}{36} = \frac{5}{18}$ **18** 0   **24** 20   **18** 0   **24** 7 (6·666)

unit 20

The pupils are first taught how to read information from given graphs (Exercise 20.1). Then how to draw a graph from given information and to obtain further information from the graph (Exercise 20.2). In Exercise 20.3 the pupils first have to complete a table of values, then plot a graph, and then obtain further information from the graph.

a 16 °C
b $17\frac{1}{2}$ °C
c Midday and 5 p.m.
d 1 a.m., 5 a.m., 11.30 p.m.
e 21 °C
f $11\frac{1}{2}$ °C
g 2.30 p.m.
h 2.30 a.m.

Exercise 20.1

1 **a** 3 cm; **b** 1 cm; **c** 3 days; **d** 6th day; **e** After 7th day; **f** $13\frac{1}{2}$ cm; **g** 11 cm; **h** 3 cm; **i** 1·7 cm; **j** 6th day.
2 **a** 400 cm; **b** 160 cm; **c** 370 cm; **d** 7 a.m. and 7 p.m.; **e** 1 a.m. and 1 p.m.; **f** 4 a.m., 10 a.m., 4 p.m., 10 p.m.; **g** Between 5 a.m. and 9 a.m., and between 5 p.m. and 9 p.m.
3 **a** 2·3 cm; **b** 1·6 cm; **c** 1·4 cm; **d** 2·6 cm; **e** 2·9 cm; **f** 1·7 sq cm; **g** 3·35 sq cm; **h** 2·5 sq cm; **i** 4·5 sq cm; **j** 6·6 sq cm.
4 **a** 4 mm; **b** 14 sec; **c** 4 mm; **d** After 15 sec; **e** 5 mm; **f** During the 19th and 20th sec; **g** During the 6th sec; **h** 2 secs; **i** 13 mm; **j** 1·4 mm per sec.

Exercise 20.2

1. **a** 350; **b** 450; **c** 650; **d** 1970; **e** They are only rough estimates, it is possible, although very unlikely, that in 1966 for example 1000 cars were sold; **f** 1974; **g** 800.
2. **a** 450; **b** 350; **c** 450; **d** 1974; **e** See answer to **1e**; **f** 1978; **g** 600; **h** No, it is too far into the future to be able to make predictions. **i** End of 1969 or start of 1970. Falling sales.
3. **a** 2 mins; **b** 4 mins; **c** 10 mins; **d** $5\frac{1}{2}$ mins; **e** $8\frac{1}{2}$ mins; **f** 12 kg; **g** 16 kg; **h** 13 kg; **i** $18\frac{1}{2}$ kg.
4. **a** 0·1 g; **b** 2·5 g; **c** 8·1 g; **d** 0·63 g; **e** 5·6 g; **f** 3·2 cm; **g** 7·1 cm; **h** 8·9 cm; **i** 8·7 cm; **j** 9·2 cm.
5. **a** 9·1 hours; **b** 7·1 hours; **c** 5·8 hours; **d** 4·9 hours; **e** 4·3 hours; **f** 71 km per hour; **g** 91 km per hour; **h** 107 km per hour; **i** 128 km per hour.

Exercise 20.3

1. 4 8 12 16 20; **a** 2·7 kg; **b** 6·7 kg; **c** 10·7 kg; **d** 14·7 kg; **e** 18·7 kg; **f** 2·3 m; **g** 3·8 m; **h** 6·8 m; **i** 12·8 m; **j** 14·3 m.
2. 10 20 30 40 50; **a** 13p; **b** 23p; **c** 33p; **d** 43p; **e** 47p; **f** 4·5 kg; **g** 7·5 kg; **h** 10·5 kg; **i** 1·5 kg; **j** 13·5 kg.
3. 6·2 12·4 18·6 24·8 31·0; **a** 3·1 cm; **b** 9·3 cm; **c** 15·5 cm; **d** 21·7 cm; **e** 27·9 cm; **f** 1·6 cm; **g** 2·6 cm; **h** 4·8 cm; **i** 6·1 cm; **j** 8·7 cm.
4. 0 1 4 9 16 25 36 49 64 81 100; **a** 6·3 sq cm; **b** 20 sq cm; **c** 42 sq cm; **d** 56 sq cm; **e** 72 sq cm; **f** 3·5 cm; **g** 4·5 cm; **h** 7·6 cm; **i** 8·7 cm; **j** 9·2 cm.
5. 24 16 12 8 6 4 3 2 1; **a** 8·7 hours; **b** 6·9 hours; **c** 5·3 hours; **d** 3·4 hours; **e** 2·4 hours; **f** 2·4 km per hour; **g** 3·2 km per hour; **h** 4·8 km per hour; **i** 19 km per hour; **j** 32 km per hour.

unit 21

Exercise 21.1

1. **a** 3; **b** 4; **c** 3 by 4; **d** 12; **e** 5; **f** 8; **g** Victor; **h** Brian; **i** 26; **j** 24; **k** 11; **l** Brian, 27 marks; **m** Victor, 11 marks; **n** Brian, 53 marks.

	a	b	c	d	e	f
2	4	2	4 by 2	8	8	2
3	3	4	3 by 4	12	6	none
4	1	4	1 by 4	4	6	none

answer section/page 209

Exercise 21.2

1 a $\begin{bmatrix} 7 & 9 & 12 \\ 5 & 16 & 12 \\ 14 & 6 & 8 \end{bmatrix}$ **b** $\begin{bmatrix} 6 & 1 \\ 7 & 0 \\ 2 & 3 \end{bmatrix}$ **c** $\begin{bmatrix} 12 & 18 & 21 \\ 9 & 15 & 3 \\ 12 & 0 & 18 \end{bmatrix}$

2 a $\begin{bmatrix} 8 & 11 \\ 10 & 14 \end{bmatrix}$ **b** $\begin{bmatrix} 5 & 9 & 16 \\ 9 & 15 & 11 \end{bmatrix}$ **c** $\begin{bmatrix} 14 \\ 11 \\ 13 \\ 12 \end{bmatrix}$ **d** Cannot be done.

3 a $\begin{bmatrix} 5 & 2 & 5 \\ 0 & 1 & 0 \\ 0 & 1 & 2 \end{bmatrix}$ **b** Cannot be done. **c** $\begin{bmatrix} 32 & 5 \\ 9 & 21 \end{bmatrix}$ **d** $\begin{bmatrix} ^-3 & 1 \\ ^-4 & ^-2 \\ ^-1 & ^-5 \end{bmatrix}$

4 a $\begin{bmatrix} 8 & 10 & 14 \\ 6 & 12 & 18 \\ 8 & 18 & 10 \end{bmatrix}$ **b** $\begin{bmatrix} 15 & 21 \\ 12 & 18 \\ 9 & 18 \end{bmatrix}$ **c** $\begin{bmatrix} 207 & 126 \\ 117 & 225 \\ 153 & 27 \end{bmatrix}$ **d** $\begin{bmatrix} 17 & 12 \cdot 5 & 8 \\ 6 & 0 & 8 \cdot 5 \end{bmatrix}$

5 a $\begin{bmatrix} 0 & 5 & 22 \\ 3 & 2 & 0 \\ 4 & 3 & 36 \end{bmatrix}$ **b** $\begin{bmatrix} 0 & 9 & 39 \\ 5 & 4 & 0 \\ 7 & 4 & 60 \end{bmatrix}$ **c** $\begin{bmatrix} 0 & 18 & 78 \\ 10 & 8 & 0 \\ 14 & 8 & 120 \end{bmatrix}$

6 a $\begin{bmatrix} 50 & 70 \\ 95 & 90 \\ 40 & 80 \end{bmatrix}$ **b** $\begin{bmatrix} 100 & 120 \\ 120 & 160 \\ 80 & 100 \end{bmatrix}$ **c** $\begin{bmatrix} 50 & 50 \\ 25 & 70 \\ 40 & 20 \end{bmatrix}$

7 a $\begin{bmatrix} 40 & 60 \\ 70 & 80 \\ 50 & 70 \end{bmatrix}$ **b** $\begin{bmatrix} 10 & 15 \\ 17\frac{1}{2} & 20 \\ 12\frac{1}{2} & 17\frac{1}{2} \end{bmatrix}$

page 210/answer section

Exercise 21.3

1 $\begin{bmatrix} 3 & 6 & 13 \\ 4 & 12 & 9 \end{bmatrix}$ **2** $\begin{bmatrix} 8 & 12 \\ 15 & 18 \\ 9 & 3 \end{bmatrix}$ **3** Cannot be done.   **4** $\begin{bmatrix} 1 & 2 & 3 \\ 2 & 2 & 3 \end{bmatrix}$

5 $\begin{bmatrix} 2 & 2 \\ 1 & 0 \\ 1 & 1 \end{bmatrix}$ **6** $\begin{bmatrix} 6 & 12 & 24 \\ 9 & 21 & 18 \end{bmatrix}$ **7** $\begin{bmatrix} 4 & 8 & 16 \\ 6 & 14 & 12 \end{bmatrix}$ **8** $\begin{bmatrix} 3 & 6 & 15 \\ 3 & 15 & 9 \end{bmatrix}$

9 $\begin{bmatrix} 7 & 14 & 31 \\ 9 & 29 & 21 \end{bmatrix}$ **10** $\begin{bmatrix} -1 & -2 & -3 \\ -2 & -2 & -3 \end{bmatrix}$ **11** $\begin{bmatrix} -2 & -2 \\ -1 & 0 \\ -1 & -1 \end{bmatrix}$

12 $\begin{bmatrix} 20 & 28 \\ 32 & 36 \\ 20 & 8 \end{bmatrix}$ **13** $\begin{bmatrix} 6 & 10 \\ 14 & 18 \\ 8 & 2 \end{bmatrix}$ **14** $\begin{bmatrix} 14 & 18 \\ 18 & 18 \\ 12 & 6 \end{bmatrix}$ **15** $\begin{bmatrix} 1\cdot 5 & 2\cdot 5 \\ 3\cdot 5 & 4\cdot 5 \\ 2 & 0\cdot 5 \end{bmatrix}$

unit 22

Exercise 22.1

The following pairs are congruent.
(2), (3), (5), (8), (10), (13).

Exercise 22.2

Care should be taken here that the enlarged figures do not come off the page. The pupils should either be given large sheets of paper, or be told to restrict their scale factors to 2 or 3, preferably with the centre of enlargement close to or inside the figure.

Exercise 22.3

1 b $\begin{bmatrix} 4 & 16 \\ 4 & 4 \\ 12 & 4 \end{bmatrix}$ **2 b** $\begin{bmatrix} 9 & 3 \\ 15 & 12 \\ 9 & 15 \\ 3 & 12 \end{bmatrix}$ **3 b** $\begin{bmatrix} 6 & 6 \\ 8 & 10 \\ 10 & 14 \\ 12 & 10 \\ 14 & 6 \end{bmatrix}$ **4 b** $\begin{bmatrix} 3 & 3 \\ 6 & 3 \\ 6 & 6 \\ 3 & 6 \end{bmatrix}$

5 b $\begin{bmatrix} 2 & 2 \\ 4 & 2 \\ 4 & 4 \\ 2 & 4 \end{bmatrix}$ **7 b** $\begin{bmatrix} -1 & -4 \\ -2 & -5 \\ -3 & -4 \\ -2 & -1 \end{bmatrix}$ **8 b** $\begin{bmatrix} -2 & -8 \\ -4 & -10 \\ -6 & -8 \\ -4 & -2 \end{bmatrix}$

Exercise 22.5

1 $W=14$ cm, $L=24$ cm
2 $W=21$ cm, $L=36$ cm
3 $W=49$ cm, $L=84$ cm
4 $W=63$ cm, $L=108$ cm
5 $W=84$ cm, $L=144$ cm
6 $W=91$ cm, $L=156$ cm
7 $W=17.5$ cm, $L=30$ cm
8 $W=24.5$ cm, $L=42$ cm
9 $W=26.25$ cm, $L=45$ cm
10 $W=94.5$ cm, $L=162$ cm
11 $L=60$ cm, $W=32$ cm, $H=24$ cm
12 $L=105$ cm, $W=56$ cm, $H=42$ cm
13 $L=7.5$ cm, $W=4$ cm, $H=3$ cm
14 $L=4.5$ cm, $W=2.4$ cm, $H=1.8$ cm
15 $L=10.5$ cm, $W=5.6$ cm, $H=4.2$ cm
16 3, $a=12$ cm
17 2, $a=6$ cm, $b=10$ cm
18 4, $a=16$ cm, $b=4$ cm
19 3, $a=6$ cm, $b=12$ cm
20 9, $a=27$ cm, $b=28.8$ cm
21 2.5, $a=15$ cm, $b=12.5$ cm, $c=17.5$ cm
22 3.25, $a=19.5$ cm, $b=16.25$ cm, $c=22.75$ cm
23 1.5, $a=9$ cm, $b=7.5$ cm, $c=10.5$ cm
24 0.5, $a=3$ cm, $b=2.5$ cm, $c=3.5$ cm
25 0.25, $a=1.5$ cm, $b=1.25$ cm, $c=1.75$ cm
26 **a** 780 cm; **b** 7.8 m
27 **a** 300 cm, 200 cm, 150 cm; **b** 30 cm, 20 cm, 15 cm
28 **a** 600 m by 800 m; **b** 10 cm by 25 cm
29 9, 18 m
30 8.333, 16.66 m

Thomas Nelson and Sons Ltd
Lincoln Way Windmill Road Sunbury-on-Thames Middlesex
TW16 7HP
P.O. Box 73146 Nairobi Kenya

Thomas Nelson (Australia) Ltd
19–39 Jeffcott Street West Melbourne Victoria 3003

Thomas Nelson and Sons (Canada) Ltd
81 Curlew Drive Don Mills Ontario

Thomas Nelson (Nigeria) Ltd
8 Ilupeju Bypass PMB 1303 Ikeja Lagos

First published 1976
Reprinted 1976, 1977
© F. C. Boyde, R. A. Court, A. M. Court, J. C. Hawdon, 1976
ISBN 0 17 431007 2

Design: David and Amanda Prout
Diagrams: Illustra Design

All Rights Reserved. No part of this publication may be
reproduced, stored in a retrieval system, or transmitted, in any
form or by any means, electronic, mechanical, photocopying,
recording or otherwise, without the prior permission of
Thomas Nelson and Sons Limited.

Printed in Hong Kong